ERRATA

Plate XXIX and Figure 70 attributed to Anhydro A/S should be attributed to Niro Atomizer Ltd.

Tower-type spray dryer installation. *Courtesy Niro Atomizer Ltd.*

CHEMICAL AND PROCESS ENGINEERING SERIES
Series editor: I. L. Hepner

INDUSTRIAL DRYING

A. Williams-Gardner

Leonard Hill
LONDON
1971

An Intertext Publisher

First published in Great Britain 1971 by
Leonard Hill Books
A Division of International Textbook Co., Ltd.,
158 Buckingham Palace Road,
London, S.W.1.

ISBN 0 249 44037 7

Filmset by Photoprint Plates Ltd., Rayleigh, Essex, and printed in Great Britain by Billing & Sons Ltd., Guildford and London.

Contents

To Lance A. Mitchell who first introduced the author to this interesting subject and whose name has for many years been synonymous with Industrial Dryers.

List of Figures

List of Plates

Preface

Most textbooks on chemical engineering devote a chapter or section to the unit operation of drying; additionally, articles and papers on the subject generally or on some specific aspects of drying or types of dryer can be found in many technical and scientific journals. A quarter of a century associated with the design and application of a majority of the many types of dryer in use today in industry has suggested to the author the advantage and, it is hoped, the usefulness of a reference book devoted completely to this important subject. Drying problems occur throughout practically all industries and the chemical and process engineer is unlikely to escape the necessity for resolving one problem or another in this field, possibly on frequent occasions.

Whereas the drying of liquids and gases can be satisfactorily carried out by a very limited number of methods, and hence types of drying equipment, the removal of moisture from solids is possible, to a varying degree of efficiency, by the use of one or more of a multitude of designs of dryers made available by manufacturers who are expert in a very limited number of types and designs. Wherever possible it is only natural that they should recommend designs of their own manufacture to resolve a particular problem. Very few manufacturers have expert experience with a wide variety of systems and designs and their judgment may be inclined to be biased in a particular instance. The purchaser and ultimate user of the drying installation can be more confident in the wisdom of his capital expenditure if he can rely on sound technical assessment of the problem by his engineering staff and correct selection of suitable drying plant to meet requirements. A proper appreciation of the function and operation of the selected dryer will ensure its efficient use and application.

This book does no more than touch on the theoretical aspects of the subject, which are extensively dealt with in the literature, other than to refer to an interpretation of the mechanism of drying solids as generally accepted today. It deals with the subject essentially from the practical standpoint in the hope that it will serve as a good reference book for students of chemical engineering and engineers in industry. Chapter I describes the importance and widespread occurence of drying in industry, whilst Chapters II and III deal with theoretical considerations in the drying of solids and the classification of the many types of drying equipment available, respectively. Chapter IV is devoted to the question of selection of the most suitable type of dryer with which to resolve an industrial drying problem. Chapters V to XII inclusive deal with the individual systems and types of dryer used in the production of a dried solid from wet feed-stock. Chapter XIII describes the special cases of Freeze Drying, Infra Red Drying and Di-electric Drying all of which have made rapid progress in recent years. Chapter XIV is devoted, necessarily briefly, to the removal of moisture from gases by methods other than the application of heat, on which all other systems of drying depend.

Twenty or 30 years ago the availability and appreciation of qualified chemical engineers in industry was limited; as a result of increased output from Universities and Technical Colleges on the one hand, and advances in process technologies on the other, this no longer holds. The vendor of drying plant today has to deal with customers employing men with trained minds in technical subjects, and the purchaser for his part requires to discuss his problems with an equally well trained technical representative. Only on this basis can an intelligent and impartial decision be taken as to what plant to install. Many years of activity in this field have impressed upon the author the importance to the purchaser that he is not discussing his problem with just another salesman, but with a technical representative in the true sense who has an adequate knowledge of his subject rather than his equipment and who can fully analyse his client's problem from the technical and process standpoints. To this end, it is hoped that this book will be of value to drying equipment manufacturers. The enthusiastic and ready co-operation which the author has received in the preparation of this book, not only from his former associates in business but also his erstwhile competitors, is, he feels, confirmation of this.

Acknowledgments

The writing of this book has been greatly assisted by the facilities and services which the Board of Directors of L. A. Mitchell Limited, Manchester, have placed at the disposal of the author and which are gratefully acknowledged. His thanks are also due to the industry and patience of Miss Doris Hazlehurst, who typed out the considerable and difficult manuscript and assisted in the checking of the proofs. Particular appreciation is expressed for the assistance given by Mr. Ralph Reeve, President, and Mr. E. M. Cooke, Director of Public Relations, Bowen Engineering Inc., in connection with the chapter on Spray Drying; to Mr. Robt. W. Jeffrey, Vice-President R/D, Fostoria-Fannon Inc., for the considerable data used in the chapter on Infra red Drying; and to Mr. C. M. Leveson-Gower, Joint Managing Director, and Mr. K. L. Buckley, The Spooner Dryer and Engineering Co. Ltd., for their assistance in connection with the section on Cylinder Dryers.

The author wishes to thank the following companies who most kindly provided data, drawings and/or photographs:

A.E.I.–Birlec Ltd., Aldridge, Staffs, England.
A.P.V.–Kestner Ltd., Greenhithe, Kent, England.
Anhydro A/S, Copenhagen, Denmark.
Banner Industries Inc. Patterson Division, East Liverpool, Ohio, U.S.A.
Bowen Engineering Inc., North Branch, New Jersey.
Buell Ltd., London, England.
Buttner–Werke, Krefeld–Ürdingen, West Germany.
Calmic Engineering Ltd., Crewe, England.
Constantin (Engineers) Ltd., Norbury, London, England.
J. P. Devine Mfg. Co. Pittsburg, Pa., U.S.A.
Dunford & Elliott Process Engineering Ltd., Surbiton, Surrey, England.
Edgar–Allen–Aerex Ltd., Sheffield 9, England.
Fostoria–Fannon Inc., Fostoria, Ohio, U.S.A.

Fluostatic Ltd., Borough Green, Kent, England.
FECO, A Division of Bangor Punta Operations Inc., Cleveland, Ohio, U.S.A.
W. C. Holmes Ltd., Turnbridge, Huddersfield, England.
Head Wrightson, Stockton, Ltd., Stockton-on-Tees, England.
Hardinge & Co. of Koppers Co. Inc., York., Pa., U.S.A.
Alfred Herbert Ltd., Coventry, Warwickshire, England.
Hunt Heat Exchangers Ltd., Middleton, Lancs, England.
International Combustion Products Ltd., London, England.
Krauss–Maffei–Imperial, G.m.b.H., München, West Germany.
Link–Belt S.A., Geneva, Switzerland.
A.P.V.–Mitchell (Dryers) Ltd., Carlisle, England.
Niro Atomizer Ltd., Copenhagen, Denmark.
Pennsalt Ltd., Camberley, Surrey, England.
Proctor & Schwartz Inc., Philadelphia, Pa., U.S.A.
Radyne Ltd., Wokingham, Berks., England.
Rietz Manufacturing Co., Santa Rosa, California, U.S.A.
Rosin Engineering Co. Ltd., London, England.
The Spooner Dryer & Engineering Co. Ltd., Ilkley, Yorks, England.
Stord–Bartz Industri A/S, Bergen, Norway.
Wyssmond Co. Inc., Fort Lee, New Jersey, U.S.A.

Symbols Used in the Text

A	Total heat-transfer area	ft^2
A_e	Effective heat transfer area	ft^2
c_p	Specific heat at constant pressure	Btu/lb dry air/deg F
c_s	Humid heat	Btu/lb dry air/deg F
D, d	Diameter	ft
d_s	Bulk density of bone-dry solid	lb/ft^3
E	Absorptivity	
F	Flow rate of wet feedstock	lb/h
G	Mass velocity	$lb/h/ft^2$
H	Absolute humidity	lb moisture/lb bone-dry air
H_r	Percentage relative humidity	
H_s	Saturation humidity	
H_A	Percentage absolute humidity	
h_c	Convection heat-transfer coefficient	$Btu/h/ft^2/deg\ F$
k	Heat-transfer coefficient	$Btu/h/ft^2/deg\ F$
L	Thickness of solids bed; length	ft
L_t	Length of one heat-transfer unit	ft
M	Flow rate	lb/h
N_t	Number of heat-transfer units	
n	Rate of rotation	rev/min
P	Pressure	
p_s	Vapour pressure of water at temperature t_s :	particle density lb/ft^3
p	Partial pressure of water at temperature t	
Q	Overall heat transferred	Btu
q	Overall heat-transfer rate	Btu/h
T	Absolute temperature	°K or °R
t	Temperature (also θ)	°F or °C

t_a	Dry-bulb temperature	°F
t_w	Wet-bulb temperature	°F
$(\Delta t)_m$	Overall mean temperature difference	°F
U	Overall heat-transfer coefficient	$Btu/h/ft^2/deg\ F$
U_a	Volumetric heat-transfer coefficient	$Btu/h/ft^3/deg\ F$ on dryer volume
U_s	Surface heat-transfer coefficient	$Btu/h/ft^2/deg\ F$
V_d	Specific volume of bone-dry air	ft^3/lb
V_s	Specific volume of saturated air	ft^3/lb
v	Velocity	ft/s; ft/m
v_f	Fluidising velocity	ft/s
W	Moisture content	
W_w	Moisture content, wet-weight basis	
W_d	Moisture content, dry-weight basis	
δ	Differential	
ε	Emissivity	
θ	Time of drying	
λ	Latent heat of vaporisation	Btu/lb

CHAPTER I

The Importance and Scope of Drying in Industry

Of all the unit operations in chemical engineering that of drying has probably the widest application in industry, and a proper appreciation of the mechanism of drying and the equipment available to resolve a drying problem on an industrial scale is correspondingly important. Most manufacturing operations, particularly in the chemical and metallurgical fields, entail one or more stages where drying in one form or another is carried out. Even purely mechanical manufacturing processes may require drying of the product before it is ready for the market. It is possible that about $1\frac{1}{2}$–2% of the total annual investment in new equipment in the chemical process industries is devoted to the installation of drying equipment.

As in other technologies, great advances have taken place in the practice of drying since the First World War, if only from economic considerations. Improved manufacturing methods have resulted in greatly increased production rates and the production-line bottle-necks created by extended drying times inherent in the use of old drying methods can no longer be tolerated. Factory floor space has to be put to the fullest possible and most profitable use, and the high cost of labour demands processing methods employing a minimum of labour. High fuel costs call for maximum thermal efficiencies and high product specifications demand scientifically controlled process operations. The old methods of drying under near natural conditions in hot rooms or on hot hearths and stillages, involving long drying times, considerable manhandling and factory space and little or no control of drying conditions, are today completely unacceptable. One or two typical examples will illustrate the importance of these considerations.

The traditional method of drying pottery tableware, which is made largely from plastic clay formed on a plaster-of-Paris mould, was to use juvenile labour to carry the ware on its mould into a heated room or enclosure where it was allowed to dry slowly, often over a period of 12 hours to 24 hours. This meant that capital was tied up, in the form of clay body and moulds, for an excessive length of time; juvenile labour, now no longer available, had to enter hot humid atmospheres, often dust-laden through treading on dried clay scrap, leading to danger from silicosis, and with improved methods of making the ware to a serious bottleneck in the production line. The situation today is completely changed. Drying times under controlled conditions have been reduced to a few minutes for cups and 2 or 3 hours for large dinner plates, resulting in great reductions in the number of moulds in circulation, greatly improved working conditions, reduced losses and the feasibility of high-speed production lines. Similarly, in the Cornish china clay industry, it was the practice to pump the refined and thickened clay slurry on to vast heated floors where it was allowed to dry in a thick bed over prolonged periods, the dried clay being later dug out by hand, under extremely bad conditions of heat and dust, by a team of men. Present-day labour is not prepared to work under these conditions and demands on productive capacity cannot tolerate the slow inefficient methods involved. Although the transition to up-to-date procedures is not yet complete, most progressive china clay manufacturers employ efficient, almost completely continuous methods of producing the dried clay ready for shipment from the slurry, involving little or no labour other than supervision. Special purpose refined china clay is today produced in its finally dried condition direct from the thickened slurry in a single stage through the use of spray dryers, whilst general run-of-the-mill clays are dewatered in quick-acting, labour saving filter presses, the filter cakes from which pass mechanically to continuously operated dryers, involving no labour requirements other than supervision and maintenance. Plants can be scientifically laid out and operated at maximum efficiency whilst special shipments can be promptly dealt with on quick deliveries.

The evolution of present-day drying techniques has, of course, been a gradual one. In its earliest stages, it relied on developing improved drying methods applied to batches of product in intermittently operated dryers. Even today such methods and equipment are adequate, although possibly susceptible to improvement, for relatively small productions of special products and where a wide range of products is manufactured, as in the pharmaceutical, dyestuffs and pigment industries. The general tendency, however, for a variety of reasons, is towards large tonnage productions of a restricted range of products calling for large production units and continuous material flows. It is in this direction that the greatest advances have been made in industrial dryer designs. The huge production of modern detergents and certain pigments at low cost would not be possible without suitably designed drying equipment capable of dealing with the quantities involved efficiently, compactly and with an absolute minimum of labour whilst still maintaining final product quality. The same applies to the metal finishing industry as an adjunct, for example, to motor car manufacture, and to the paper, decorative laminates, veneer and fibre board industries. The accent is on rapid but correct drying occupying a minimum of factory space, employing a minimum of labour, and the highest possible thermal efficiency commensurate with product specification guarantees.

The term "drying" is difficult to define within rigid limits. From the industrial standpoint it is understood to represent the removal of a liquid—usually, but not always, water—from a solid by thermal means. It can also mean the removal of water or other volatile liquid from another liquid or gas, or the removal of water from a suspension or solution of a solid. Such a definition clearly distinguishes thermal drying from mechanical dewatering as in the action of filters or centrifuges. It does not, however, distinguish drying from the process usually referred to as evaporation as carried out in commercial evaporators. In evaporators heat is employed to remove liquid from solid solutions or suspensions in the liquid, the liquid generally occurring in larger proportions than occur in the processes of drying. However, as will be seen, spray dryers and film drum dryers are employed to recover a dried solid product from an initial feed stock having a preponderating content of water or other volatile liquid component. The most satisfactory method of differentiating between drying and evaporating is on the basis of the type of equipment employed in the two processes and that in evaporational processes much larger quantities of liquid component are recovered per hour and per unit of plant than occurs in drying operations. Moreover, the solid component is recovered from an evaporator either in the form of a concentrated solution or suspension or as a wet slurry, whereas that from a dryer is either quite dry or substantially so. There are also the special cases of drying when a final product in a dehydrated or anhydrous condition is arrived at, as for instance in the removal of moisture from "live" foodstuffs in the case of the former or the removal of water of crystallisation from inorganic salts or hydroxyl groups from organic compounds in the case of the latter.

There are many reasons, either technical or commercial, why the process of drying is essential in the production of a final product. The product must be in suitable form either for subsequent processing or for marketing. The clay granules which are fed to a press as, for instance, in tile manufacture or the production of low tension insulators must flow satisfactorily in the presses and the most suitable characteristics occur at a certain moisture content of the clay. A cup prepared in a plaster-of-Paris mould must be of the correct moisture content for subsequent sponging or turning and for fitting the handle. Pharmaceutical granules require to be predried for feeding to the tableting machine. Dyestuffs and pigments, foodstuffs, cereals and detergents require to be dried to the correct final moisture content ready for packaging. Physical suitability of the final product for the market may be controlled by correct drying methods, as in the case of globular detergent powder of uniform size and appearance, or synthetic fertilisers of granular form to minimise dust losses at the time of application; pigments have usually to be offered on the market in a fine state of subdivision and adequate drying is imperative so that the pulverising equipment can handle the product satisfactorily.

Transportation costs are usually directly related to the moisture content of the product and a balance must be struck between the cost of transporting moisture on the one hand and the cost of drying on the other. For example, coal slurries would be virtually valueless if not dewatered and to some extent dried before shipment; the same applies to ore concentrates from mining and metallurgical processes.

In the absence of drying operations, subsequent processing can become uneconomic, as for instance in the ceramic industry where predrying in less expensive plant ahead of firing kilns of high capital cost can make the overall economics of the process much more attractive. Similarly the value of by-products can be greatly enhanced by introducing efficient drying operations, as in the processing of poultry manure in conjunction with large batteries; blood-drying plants in slaughterhouses; calcium carbonate from water-softening plants; spent grain from distilleries; bone meal from gelatine plants and wood meal from timber plants, to quote a few examples.

The unit operation of drying as applied to solids is by definition a thermal process and as such is expensive in energy depending on the amount of water to be removed. Other than in exceptional cases, it is more economical to remove as much water from the solid by mechanical separation as a pretreatment or dewatering stage. In the case of suspensions of solid in water this will be carried out by filters and in the case of very wet unpumpable solids by means of hydro-extractors, the resultant dewatered solids passing to a thermal drying unit. Special considerations may constitute a case for direct thermal drying of a suspension of solids in water. The elimination of all labour or a particularly desirable physical form for the dried product may justify the cost of the thermal energy required to remove the water content which could otherwise be reduced more economically by mechanical means. Thus, whereas clay tile body in suspension in water can be dewatered in filter presses and the filter cake subsequently dried in a suitable dryer followed by a granulating unit prior to feeding to the tile presses, all intermediate stages can be eliminated and the clay suspension fed directly to a spray dryer which yields the product in an admirable condition for feeding to the tile presses. Filter presses, conveyors and granulating plant together with the necessary drying plant have thus been replaced by a single operation, namely spray drying, with a completely suitable end-product. Considerable savings in factory floor space, labour and power together possibly with some economy of capital outlay will have been effected. In general, however, a case can usually be made out for preliminary mechanical dewatering ahead of the actual thermal drying operation. This sequence of operations is almost universally employed in the pigment, dyestuffs, clay and diatomaceous earth, heavy chemicals and many other industries. Depending upon the degree of preliminary dewatering desired and the ease or otherwise of mechanical removal of excess moisture, plate-and-frame or continuous rotary filters are employed.

Certain materials have the capacity for absorption of considerable quantities of water through capillary attraction by virtue of their porous nature, and advantage is sometimes taken of this to accomplish a measure of preliminary dewatering. Frequently this is resorted to in order to present the product in a physical form of adequate mechanical strength to withstand a subsequent and possibly final thermal drying operation. Plaster-of-Paris has a strong attraction for water through capillary action and it is customary to dewater a clay cream or "slip" in the casting operation of certain pottery in order to produce an article in the correct physical form and having adequate mechanical strength to withstand subsequent finishing operations and final thermal drying. The thickness of the article in its dewatered state will depend on the length of time the "slip" is left in contact with the plaster moulds, the excess slip being poured off. It is of course necessary to remove the absorbed water by thermal drying of the mould before re-use. Similarly in the manufacture of lozenges, gums, jelly babies and similar types of confectionery, the concentrated and usually hot syrup is fed into starch moulds which absorb a material proportion of the contained moisture, the remainder being removed by thermal drying whilst the confectionery is still supported in the starch mould. Again the starch must be redried to recover its absorptive power for water.

Drying, as distinct from preliminary dewatering, may also be accomplished by adsorption. This usually applies to the drying of gases and would constitute a complete drying operation. Certain solids, such as activated alumina, silica gel and carbon have a propensity through physico-chemical attraction within their structure of adsorbing high concentrations of fluids whether liquid or gaseous, and advantage is taken of this property to remove moisture from gases. Such materials have a limited but known capacity for moisture adsorption and can be re-activated by the action of heat. Quite apart from their application to the total drying of a gas, they may be used to reduce the humidity of air employed in thermal drying operations, especially where these have to be carried out at relatively low temperatures as in the drying of gelatine.

A further method of drying both gases and liquids, especially where small traces of moisture have to be removed, is by the use of chemical desiccants such as metallic sodium, quicklime, calcium chloride, phosphorus pentoxide and anhydrous salts. This method of drying is of very limited application and is outside the scope of this book.

A technique of drying developed in recent years and which is attaining increasing importance is that of freeze drying. This is to be distinguished from simple refrigeration as applied to gases or liquids when moisture is removed by condensation to water or even to ice as in the practice of air conditioning. True "freeze drying" essentially consists of freezing the wet material and through the application of mild and controlled amounts of heat under high vacuum subliming the moisture from the solid ice state direct to water vapour without passing through the liquid phase.

This technique has attained considerable importance in the drying of extremely heat-sensitive materials such as blood plasma, antibiotics and materials the original flavour and perfume of which must be retained unimpaired in the drying product.

In any discussion of drying, certain terms are peculiar to the subject and are in general use; it will be helpful to define these at the outset.

Definition of Terms Employed

DRYING is generally interpreted as meaning the removal of moisture composed of water although in special cases a liquid other than water may be involved. Considerations which apply to the removal of water in a drying operation apply equally to other liquids, subject only to the effect of different values of such properties as boiling point, specific heat, latent heat of vaporisation, vapour pressure and molecular weight.

MOISTURE CONTENT of a solid is usually expressed as the moisture content by weight per unit weight of the dry or wet solid. Moisture content on a volumetric basis is rarely if ever employed and throughout this book will refer to the weight basis. It is customary in the United Kingdom to refer to moisture content on the wet-weight basis but the more scientific basis of dry weight is being increasingly adopted in dryer calculations. It is important to understand the significance of the two bases and to be quite clear as to which ratio is provided when embarking on dryer design calculations. Wet-weight basis expresses the moisture in a material as a percentage of the weight of the wet solid, whereas the dry-weight basis expresses the moisture in a material as a percentage of the weight of bone-dry material. Thus 50% moisture on the wet-weight basis (w/w basis) signifies 50 parts by weight of water per 100 parts by weight of wet material. On the other hand, a material containing 50% of moisture on the dry-weight basis will contain 50 parts by weight of water per 100 parts by weight of bone-dry material. The former converted to the dry-weight basis would mean a moisture content of 100% on the bone-dry (b/d) basis whilst the latter represents 33·3% on the w/w basis; the relationship between wet-weight and dry-weight basis can be expressed as

$$W_w = \frac{W_d}{1+W_d} \qquad W_d = \frac{W_w}{1-W_w}$$

where W_w is the pounds of moisture per pound of wet material and W_d is the weight of moisture per pound of bone-dry material. This relationship is shown graphically in Fig. 1.

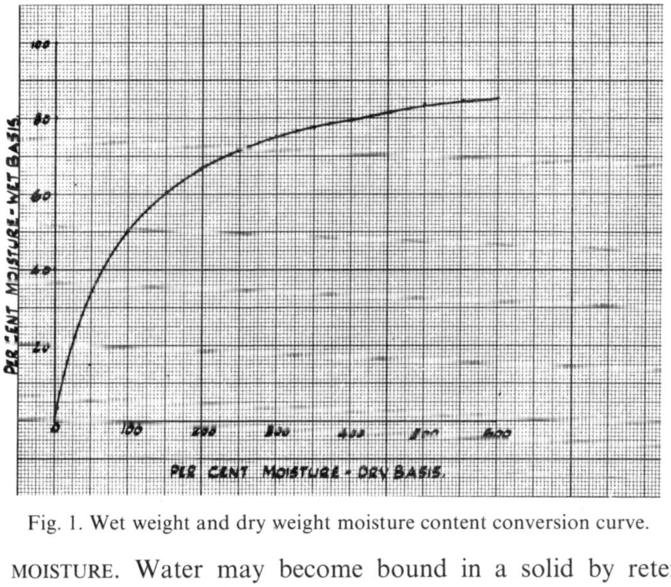

Fig. 1. Wet weight and dry weight moisture content conversion curve.

BOUND MOISTURE. Water may become bound in a solid by retention in capillaries, solution in cellular structures, solution within the solid or by chemical or physical adsorption on the surface of the solid. This water will exert a vapour pressure lower than that of pure water at the existing temperature and can only be removed from the solid under specific conditions of humidity in the surrounding atmosphere.

UNBOUND MOISTURE in a hygroscopic material is the moisture in excess of the equilibrium moisture content corresponding to the saturation humidity. All the moisture content of a non-hygroscopic material is unbound moisture.

FREE MOISTURE CONTENT is the moisture content which is removable at a given temperature and may include both bound and unbound moisture.

EQUILIBRIUM MOISTURE CONTENT is that to which a material can be dried under specified conditions of temperature and humidity of the surrounding atmosphere.

CONSTANT RATE DRYING PERIOD is that period in a drying operation during which the rate of moisture removal per unit of drying surface is constant.

CRITICAL MOISTURE CONTENT is the moisture content existing in the material at the termination of the constant rate drying period.

FALLING RATE DRYING PERIOD is that period of the drying cycle during which the instantaneous drying rate continuously decreases and which commences at the end of the constant rate period.

INTERNAL DIFFUSION is the movement of the moisture liquid phase through the mass of solid during the drying operation.

DRYER EFFICIENCY is that fraction of the total heat supplied during a drying operation which is usefully used in evaporating moisture.

BATCH DRYING is an intermittent drying operation where a batch of material is charged into a dryer and the drying completed before its removal from the dryer.

CONTINUOUS DRYING constitutes a continuous feed of moist material to a dryer and the continuous discharge from the dryer of the dried material.

SEMI-CONTINUOUS DRYING occurs when batches of material are intermittently fed to the dryer and equal batches of dried material are simultaneously removed from the dryer.

COUNTER-FLOW DRYING occurs when the drying medium and product move in opposite directions to one another within the dryer.

PARALLEL-FLOW DRYING (sometimes referred to as CO-CURRENT FLOW) occurs when the drying medium and product move in the same general direction through the dryer.

DIRECT-HEAT DRYERS embrace dryers in which the drying medium is brought into direct and intimate contact with the material subjected to drying.

INDIRECT-HEATED DRYERS operate without direct contact between the heating medium and the material.

INDIRECT-DIRECT-HEATED DRYERS employ a combination of indirect and direct contact between the heating medium and the product.

THROUGH-CIRCULATION DRYERS are designed in such a manner that the heated drying medium (air) is passed through a bed of the material subjected to drying.

A fuller appreciation of the importance and scope of thermal drying in industry can be indicated by reference to the more important industries in which one or more drying operations are essential stages in the manufacture of the final product and the types of dryer employed. With very few exceptions, such as petroleum products and metals, almost all products from present-day industry undergo thermal drying so that any such reference list cannot fail to be incomplete. Many products which are dried are referred to in the chapters describing different types of dryers.

The Colour, Pigment and Dyestuffs Industry

Pigments are widely used in the manufacture of paints for surface coatings and in the ink, plastic, rubber, paper and linoleum industries. They may be derived from naturally occurring clays and earths, by the preparation of precipitated inorganic salts or by the synthesis of insoluble organic dyes. Typical examples of earth colours are ochres, sienna, umber and Vandyke brown. Some of these are gradually being replaced by pigments of greater brilliance produced chemically as are many other coloured pigments such as the various shades of blue, red, yellow and green. White pigments are made on a very considerable scale of manufacture, notably titanium dioxide, white lead and lithopone. The preparation of earth colours usually entails a certain degree of lixiviation with water entailing subsequent drying, whereas the chemically produced pigments involve a wet process requiring subsequent dewatering by filtration followed by thermal drying unless the dewatering and drying is carried out in a single operation of thermal drying. Black pigments are largely composed of carbon black, which is a by-product of the petroleum industry and, particularly in its pelletised form, requires a degree of drying before shipment. Insoluble organic dyes are used either directly as a pigment or indirectly, being precipitated on an insoluble inorganic base or co-precipitated with the inorganic base. Here again, the process is a wet one requiring a drying operation. The organic dyestuffs industry is vast and world-wide; the manufacturing processes involved are essentially wet processes and necessitate thermal drying of the final product before shipment.

Not only is thermal drying an essential stage in the manufacture of colours and dyes, but the manner in which thermal drying is carried out is also of great importance. Many inorganic colours and most organic dyes are heat-sensitive; the time–temperature effect in the drying operation may be critical in arriving at the correct shade of colour or in the elimination of thermal degradation. Until improved dryer designs and their operational techniques were evolved, it was usual to dry most organic dyes and some inorganic pigments in batch vacuum tray or shelf dryers, in spite of their inherent operational disadvantages, in order that drying could be accomplished at relatively low temperatures and in the absence of air. Modern designs of dryers, both batch and continuous, are of such relatively high efficiency, are amenable to such close control of drying conditions, and are capable of such greatly enhanced drying rates that cumbersome vacuum batch dryers are today rarely used. Without doubt, the most widely used type of dryer employed in this industry for multiple product manufacture is the recirculation batch type truck-and-tray compartment dryer, which for fairly large productions may take the form of a medium-sized tunnel dryer. Modern designs of this type of dryer, incorporating internal air recirculation and reheater coils with proper distribution of the drying air, make possible close control of drying temperatures and relatively short drying cycles such that the product meets shade and quality specifications satisfactorily. For large tonnage productions of pigments such as titanium dioxide, lithopone, white lead, oxides of iron and certain organic dyestuffs, the use of continuously operating designs of

Table 1. Application of dryers in the pigment and dyestuffs industry

Product	*Type of dryer*	
	Normal productions and multiple products	*Large productions of one product*
Earth colours and by-product from oxides	Truck-and-tray compartment Trough Agitated pan	Rotary Pneumatic
Precipitated inorganic colours	Truck-and-tray compartment Agitated pan	Continuous band Spray Rotary
Carbon black		Indirect rotary
Titanium dioxide, Lithopone, white lead, whiting		Semi-continuous truck-and-tray tunnels Continuous band Spray
China clay (as white filler)		Direct and indirect rotary Turbo-tray Continuous band Film drum
Dyestuffs	Vacuum tray Truck-and-tray compartment Film drum Agitated pan	Continuous band Spray

dryer becomes possible, resulting in greatly improved economy of manufacture. The types of dryer used today in this industry for a selection of its products are indicated in Table 1.

Pharmaceutical and Fine Chemicals Industry

This industry covers a considerable range of products, both inorganic and organic, many of which require drying before packaging. On the quantitative basis, probably the most important is the production of pharmaceuticals in tablet form, for which the product is usually produced in powder or granular form which requires to be dried from moisture in either the aqueous or organic solvent form. Until recently this was done in small air recirculation drying cabinets holding ten to 20 trays or somewhat larger truck-and-tray ovens, but the present tendency is towards batch fluidised bed dryers with much shorter drying times and no necessity for filling and emptying trays. Where a steady and continuous flow of granules or powder is produced the drying operation can be put on a continuous basis by employing simple vibratory conveyors subjected to a combination of convected and radiant heat. Where productions of a single product are considerable, such as magnesium carbonate, aspirin or similar materials, larger scale continuous dryers such as turbo-tray or continuous through-circulation dryers are employed. Excessively heat-sensitive products such as antibiotics and blood plasma call for special treatment and are usually dealt with in freeze-drying installations or high-vacuum tray or drawer type dryers. Large productions of crystalline chemicals, whether for the removal of surface free moisture or a degree of chemically bound moisture, can be successfully dealt with in pneumatic dryers. Bulk productions of some products are also dried in steam-heated continuous rotary dryers whilst smaller productions are conveniently handled in agitated pan dryers. Table 2 gives an indication of the distribution of dryer types in this industry

Natural Ores, Minerals and Heavy Chemicals

The industries engaged in the production of materials falling within this range will normally be dealing with considerable tonnages and the drying plants used will be of the types designed for large throughputs. Of these, continuous rotary dryers, either of the direct, indirect or indirect-direct types, outnumber the remainder although with increased experience and improved designs continuous fluid bed dryers are beginning to assert themselves. Both types can be designed to deal economically with very large tonnages and can be installed either in the open or with the scantiest of weather protection. Natural ores and minerals such as pyrites, coal, sand and clays would normally be dealt with in these types of dryer as they are capable of trouble-free operation over long periods with a mini-

Table 2. Applications of dryers in the pharmaceutical and fine chemical industry.

Product	*Type of dryer*
Powders and granules before tableting	Batch fluid bed dryers Compartment tray drying stoves Drawer or shelf vacuum dryers Agitated pan dryers Radiant heat vibratory conveyor dryers
Crystalline products and fine chemical powders	Pneumatic dryers Turbo-tray dryers Truck-and-tray tunnels Continuous fluid bed dryers
Precipitated fine chemicals	Filtration followed by pre-forming and continuous through-circulation band dryers
Heat sensitive products in solution	Spray dryers Vacuum film drum dryers
Highly heat sensitive materials, antibiotics, blood plasma, glandular extracts	Freeze dryers

mum of maintenance and supervision. They are also widely used for large tonnage productions of fertilisers, ammonium sulphate and metallurgical concentrates. Where the material is amenable to treatment in continuous pneumatic dryers without excessive troubles arising from abrasion, this type of dryer may also have considerable advantages, particularly when a disintegrator or pulveriser is incorporated in the system. Metallurgical concentrate filter cakes from continuous vacuum filters are also frequently dealt with, with or without pre-forming, in continuous through-circulation band dryers and turbo-tray dryers.

Paper and Allied Products

By far the largest demand for drying equipment in this field is for the continuous drying of paper and this is generally carried out on cylinder or "can" drying machines in which the paper passes continuously through the dryer at high speed in contact with steam-heated machined and ground cylinders which by conducted heat transmission, sometimes assisted by forced convection, remove the moisture from the paper. The quality of paper will range from the lightest weight material to heavy-weight papers. Wallpapers and certain other types may be dried in festoons propelled through a forced convection drying enclosure or in a single pass-through drying tunnel supported by and subjected to the action of high-velocity heated air jets. Fibre board, in cut lengths, leather board and straw board in cut lengths are usually dried in specially designed continuous conveyor dryers, the boards being subjected to high-velocity air jet systems on a recirculatory principle capable of maintaining correct temperature and humidity conditions. Moulded paper pulp articles similarly are dried in conveyorised dryers. Resin-impregnated papers for the decorative laminates market are dried after impregnation in tunnel dryers usually having a combination of a conveyor of special design together with high-velocity air jets which are operated in such a manner as to "float" the paper in its passage through the dryer so as to eliminate marking of the paper whilst still wet. As a result of increasing speeds of production of these several products and the highly competitive nature of their markets, dryer designs and performance have been greatly improved in recent years.

Table 3. Application of dryers in the paper and allied products industry

Product	*Type of dryer*
Newsprint, Kraft, wallpaper, paper tissue, paper board, grease-proof paper, paper pulp	Multiple cylinder and large single cylinder Yankee machines
Coated wallpapers	Continuous festoon dryers, high-velocity air jet tunnels
Fibre board, leather board, straw board	Conveyor band dryers, conveyor clip dryers and rotating tray dryers
Moulded paper pulp articles, containers, etc.	Continuous through-circulation band dryers continuous conveyor shelf dryers, semi-continuous truck-and-tray tunnels
Resin-impregnated papers for decorative laminates, coated papers	High velocity jet continuous tunnel dryers

Foodstuffs and Agricultural Products

Drying technology has developed in the foodstuffs and agricultural fields to a very considerable extent in recent years, no doubt accelerated by the requirement for dehydrated vegetables and other foodstuffs during the Second World War. Previously, processed potatoes, maize and other starches and milk were already dried on a large scale. War-time conditions, involving shortage of shipping, the global distribution of the fighting forces, the necessity for high packing-density products, emergency long-term food storage and the special requirements for easily digested high food content products for near-starved populations and prison camps gave considerable impetus to food-drying techniques, which have advanced still further in peace time to meet the demands of commercially promoted markets for predried products of all types in the kitchen. There exists today a vast industry which prepares and sells dried vegetables, fruits, meats, soups and beverages, the qualities of which are constantly improving.

The drying of milk, coffee and starches has today attained vast proportions and the favourable acceptance of many other products in a dehydrated form has resulted in a rapid increase in industrial drying operations in this field. Thermal drying is an essential step in the manufacture of a wide range of confectionery and a very high level of expertise is called for to meet increasing market requirements by accelerated production techniques.

The harvesting of hay and grain crops in countries such as Great Britain has always been subject to the vagaries of climate; prolonged heavy rainfall at harvest time can result in considerable losses and deterioration in crop quality. Protein and carotene values in hay may seriously diminish even if a crop is not a total loss; in any event it will be less than in fresh green grass and only one or at most two crops can be grown in a single season. With appropriate husbandry several crops of grass of the highest quality are today cut and pass to drying installations regardless of weather conditions, the final product having an extremely high food value. The increasing use of combine harvesters, the conservation and storage of seed grain and the requirements of flour-milling companies and breweries have resulted in the thermal drying of grain on a considerable scale in suitably located installations.

Table 4. Application of dryers in the food and agricultural industries.

Products	*Type of dryer*
Vegetables, confectionery, fruits, pectin	Compartment and tunnel tray
Grass, grain, vegetables, fruit, nuts, breakfast cereals	Conveyor band
Grass, grain, apple pomace, lactose, poultry manure, peat, starch (vacuum)	Rotary
Coffee, milk, tea, fruit purees	Spray
Milk, starch, predigested infant foods, soups, brewery and distillery by-products	Film drum
Starch, fruit pulp, distillery waste products, crops	Pneumatic
Coffee, essences, meat extracts, malted and other confectioneries	Freeze dryers and vacuum dryers

Ceramics Industry

Drying operations are an essential part of this vast industry, from the raw materials to the finished product. The natural or refined china and ball clays, felspar, and the other components of the clay body used in the manufacture of ceramic products require to be dried before delivery to the industry. The frits, glazes and colours are frequently produced by a wet process and may be delivered to the potter in the dried state. Domestic tableware formed from the plastic clay on the potter's wheel, whether handmade or, as now, more usually by automatic and semi-automatic methods, requires to be dried either to the "white" condition or for mould release, as in the case of cups and hollow ware; cast ware, whether for the table or for sanitary installations, is dried either in the mould for mould release or after removal from the mould when it has attained sufficient mechanical strength to be self-supporting. Glazes applied to biscuit ware for twice-fired products have to be dried after dipping or spraying before the glost-kiln. With the advent of automatic and semi-automatic jigging, jolleying and roller making machines, the rates of production of tableware have increased enormously and the space occupied, the number of moulds employed and capital tied up in materials would be prohibitive if means for accelerated drying were not available. Prolonged drying as was customary by traditional methods, involving up to 24 hours for a dinner plate for example, constitutes a most serious bottleneck on production. Considerable development work and a better understanding of the mechanism of drying of plastic clay, which will start with a moisture content of about 23% (wet-weight basis) and which undergoes considerable shrinkage in the course of drying, particularly to the "white" state, has resulted in great reductions in drying time. This, in turn, has made it possible to mechanise the drying operation to a greater or lesser extent, so eliminating the need for cheap juvenile labour – no longer available – and generally improving working conditions for personnel. Forced convection dryers with reasonable control of humidity conditions now make possible the drying of large dinner plates in one to two hours and of cups in 5–10 minutes to the "leather-hard" condition for mould-release and subsequent turning and handling operations. The latest techniques employing relatively high temperatures (up to 300°C) and high-velocity air jets with careful screening and cooling of the plaster-of-Paris mould now indicate that times as low as 20 minutes for large plates and less than half a minute for cup mould-release may be commercially possible. Glaze drying may be greatly accelerated by application of infra-red or radiant heat.

Sanitary ware and high-tension electrical insulators, many of considerable size and wide variations in thickness in a single article, present a special problem. Traditional methods of drying have involved standing the ware on warm floors or in warm rooms for several days to permit very gradual drying and shrinkage to the leather-hard condition and this has constituted a serious bottleneck on production, with large quantities of ware of considerable value occupying factory floor space, and always subject to the danger of damage from one cause or another. Carefully controlled humidity drying over predetermined temperature and humidity cycles has reduced drying times to a matter of hours and has made possible continuous or semi-continuous flows of goods through drying tunnels. Articles having differing drying times can be loaded onto separate conveyor lines in the same tunnel having zonal control of relative humidity and temperature of the drying air, so as to comply with the appropriate time–temperature–humidity cycles.

Glazed wall and floor tiles – as also low-tension electrical porcelain – are manufactured by feeding granular clay body "dust" to presses with a moisture content of the order of 7–8%. The traditional method of preparing the clay has been to filter-press the body slip, dry the filter cakes on steam-pipe heated racks, break up the dried cakes and pass them through an edge-runner type granulating mill, altogether a cumbersome method involving considerable labour, factory space, unhealthy atmospheres and poor control of moisture content in the clay granules. Improved results and conditions were secured by loading filter cakes into truck-and-tray or "toast-rack" drying tunnels, but modern practice involves the production of the dried clay at a closely controlled moisture content direct from the clay slip either on film drum dryers or in spray dryers. Some manufacturers pass specially prepared high solids content slip – similar to casting slip – direct to the dryer whilst others prefer to filter the original slip and reblunge the filter cake to the required solids content. High solids content slip of the order of 70% by weight, in order to have adequate fluidity, requires to have added about 0·2% of an electrolyte such as soda ash or sodium silicate and this is quite normal in the preparation of casting slips. The higher the solids content the greater the yield and economy of the drying plant. In the case of film drum dryer operation the product is a mixture of small flakes and dust which passes continuously to the granulating unit and thence to the tile presses. The spray-dried product, on the other hand, is in an excellent form with high flowability characteristics ready for pressing.

Textile Industry

As in the ceramic industry, thermal drying plays a most important part in the textile industry from the raw stock to the finished product. Whether derived from the sheep's fleece or cotton field or man-made fibre chemical processes, there are various stages of scouring, bleaching or dyeing where moisture is necessarily introduced and has to be removed either by mechanical, thermal or a combination of both methods. The spun stock as thread or yarn similarly passes through wet processes and will have to be dried in hank, skein or packaged form. The woven cloth may be bleached, dyed or printed by a wet process. Considerations of space, time, labour and

Table 5. Application of dryers in the ceramics industry.

Product	*Type of dryer*
TABLEWARE: FLAT	
Plastic clay on the moulds	Intermittent forced convection or jet "Dobbin" drying ovens
	Continuous forced convection or jet conveyor dryers
Glaze dipped or sprayed ware	Continuous forced convection or jet conveyor dryers. Infra-red dryers
HOLLOWWARE	
Ware in the mould	"Dobbin" ovens
	Conveyorised forced convection and jet shelf dryers
SANITARY WARE	
"Green" ware after removal from the mould	Hot floor and hot room drying
	Conveyorised forced convection humidity controlled drying tunnels
Dipped or sprayed glazed ware	Conveyorised forced convection tunnels
ELECTRICAL PORCELAIN:	
Low-tension	Compartment and conveyorised forced convection dryers
High-tension	Warm rooms
	Continuous and semi-continuous humidity controlled tunnels
GLAZED TILES:	
Tile body	Filter cake drying in tunnels
	Filter cake extrusion and continuous through-circulation band dryers
	Direct drying of prepared clay slip on film drum dryers or in spray dryers

uniform quality of product demand efficient designs and operation of drying equipment specifically designed to carry out drying in an automatic manner involving a minimum of labour and factory floor space.

Loose wool, whether at the fellmonger's works or the woollen mill, and loose cotton have for many years been dried continuously on through-circulation band dryers of one form or another, the most modern being of the conventional straight-through single-deck type, preceded by mechanical openers, a measure of preliminary dewatering having been carried out by hydro-extractors or squeeze rolls. The openers and mechanical feeders ensure a uniform spread of stock on the conveyor band to promote uniform drying and avoidance of damp patches. A more recent development for loose fibres after scouring, bleaching or dyeing has been the use of a series of perforated rotating drums on one half of the surface of which, alternately upper and lower, a layer of opened stock is formed by suction, the stock being transferred from one drum to the next from feed point to discharge from the dryer. Heated air on the recirculation principle is sucked through the stock at high velocity promoting rapid and uniform drying. Yarn in hank, tow, sliver and tubular knitted fabrics may also be dried in the above types of dryer. Other materials similarly dried are rayon staple fibre, cellulose acetate, man-made fibres generally and sisal.

Yarn in hank, both cotton and wool, and artificial silk in skeins are dried in either compartment or continuous conveyorised dryers suspended on horizontal poles, which may be wood or extruded aluminium section, drying being by means of hot air circulation downwards through the suspended materials. In the case of conveyorised dryers, usual for large throughputs, the poles are arranged to rotate on their axes in order to avoid damp patches. It is not unusual in such dryers to include a conditioning section so that the finally delivered yarn has the requisite and permissible amount of retained moisture or "regain". Yarn in packaged form as cheese, cake, pirn or cop is dried in compartment, tunnel or conveyorised shelf dryers, the drying conditions of temperature, humidity and air movement being such as to yield uniformly dried product of satisfactory quality in the minimum of time. In the case of pressure-dried cheeses and cakes, special designs of moderately high-pressure drying cabinets or ovens are used.

Piece goods after bleaching, dyeing or printing are dried in a variety of designs of continuous dryers depending on whether the material has to be processed in a tensionless or stretched manner. Festoon or loop dryers are widely used where the material has to be dried in the absence of tension. In these, the continuous feed of material is automatically draped in loops on supporting poles carried at their ends on conveyor chains, and is subjected to hot forced convection currents of air in a downward direction. On discharge from the dryer, the material is automatically removed from the poles and folded into receiver bins. Modern designs of dryer employ high velocity hot air jets impinging on upper and lower surfaces of the material as it moves in a single pass through a drying tunnel, the material being "floated" on the lower air jets. Very rapid drying is possible by this method making possible drying tunnels of reasonable length. Tenter dryers are used for drying piece goods under tension or stretched. The material is stretched in a horizontal direction between two endless conveyor chains equipped with clips or pins which locate and grip the edges of the fabric as it enters the dryer. Here again, the modern practice is to employ high velocity hot air jets impinging vertically onto the surfaces of the fabric to promote accelerated drying.

Table 6. Applications of dryers in the textile industry

Material	*Type of dryer*
Fellmongers wool, scoured, bleached and dyed wool or cotton	Continuous through-circulation band dryers
Cellulose acetate, man-made fibres, sisal, tow, sliver and tubular knitted fabrics	Continuous suction drum dryers
Yarn in hank form	Compartment ovens Continuous pole tunnels Through-circulation band dryers
Yarn in package form: cheeses, cakes, cops, pirns, etc.	Compartment ovens Continuous conveyorised shelf dryers
Piece goods in continuous sheets	Tensionless: Festoon, loop, jet "floating" dryers Under tension: Tenters, clip and pin Multi-pass roll dryers
Carpets	Continuous converyorised forced circulation or high velocity jet dryers
General	Forced convection ovens Conveyorised suspension dryers

CHAPTER II

Theoretical Considerations: The Mechanism of Drying

As this book is intended to be essentially a practical treatment of the subject of industrial drying as a unit operation, a detailed discussion of the theoretical side of the subject should be outside its scope. A considerable literature exists and the reader who wishes to make a further study of this aspect of the subject is referred to the bibliography at the end of the book. However, for a proper appreciation of what is involved in drying operations and to assist in a practical approach to drying problems, certain basic theories and concepts of the mechanism of drying should be understood.

The Drying of Solids

When a wet solid is subjected to thermal drying two processes occur simultaneously, namely:

1. Transfer of heat to raise the wet solid temperature and to evaporate the moisture content.
2. Transfer of mass in the form of internal moisture to the surface of the solid and its subsequent evaporation.

The rate at which drying is accomplished is governed by the rate at which these two processes proceed.

Heat transfer in the drying operation will occur through flow of heat as a result of convection, conduction or radiation and in some cases as a result of a combination of any of these effects. It will be seen later that industrial dryers differ in type and design dependent on the principal method of heat transfer employed but by whichever type heat requires to be transferred to the surface of the solid and thence to its interior. The one exception is that of a dielectric or high frequency electric dryer in which heat is generated internally within the solid and flows to the exterior surfaces.

Mass transfer in the drying of a wet solid will depend on two mechanisms:

movement of moisture internally within the solid which will be a function of the internal physical nature of the solid and its moisture content; and the movement of water vapour from the material surface as a result of external conditions of temperature, air humidity and flow, area of exposed surface and supernatant pressure. In a drying operation either of these mechanisms may be a limiting factor on the rate of drying although they will both be proceeding simultaneously throughout the drying cycle. With some materials and at a certain stage of the drying operation, the rate of movement of moisture within the solid to the surface may be the controlling factor, whereas at some stage of the drying cycle the rate of surface evaporation may be the controlling factor. It is important to bear this in mind and to understand the processes concerned in one's approach to a drying problem and to obtain optimum conditions for a drying operation.

Internal Conditions

As a result of heat transfer to a wet solid a temperature gradient develops from the heated surface inwards while evaporation occurs at the surface. This produces a migration of moisture from within the solid to the surface which will occur through one or more mechanisms, namely, diffusion, capillary flow, internal pressures set up by shrinkage during drying and, in the case of indirect (conduction) dryers, through repeated vaporisation and recondensation of moisture progressively to the exposed surface. These phenomena may occur simultaneously and one or more may be predominant at different stages of drying but the net result will be a moisture gradient through the thickness of material. In the case of a solid which is exposed to the heat-transfer medium on more than one surface there will exist more or less identical gradients from the centre outwards to the surfaces. These moisture gradients are due to the flow of liquid from within the solid to replace moisture evaporated from the surface, and in the case of liquid diffusion is found to decrease in rate with decrease in moisture content. An appreciation of this internal movement of moisture is important in two respects in practice. Where it is a controlling factor, as occurs after the critical moisture content in a drying operation carried out to very low final moisture contents, those variables, such as air quantities, which normally enhance the rate of surface evaporation become of decreasing importance except in so far as they promote higher heat-transfer rates. Longer residence times and, where permissible, higher temperatures become necessary. Similarly, in the case of materials such as ceramics, timber, etc., where considerable shrinkage occurs, excessive surface evaporation sets up high moisture gradients from the interior towards the surface, which is liable to cause overdrying, excessive shrinkage and consequently high tension resulting in cracking or warping. In such cases it is essential not to incur too high moisture gradients by retarding surface evaporation through the employment of high air humidities whilst maintaining the highest safe rate of internal moisture movement by virtue of heat transmission. The temperature gradient set up in the solid will also create a vapour-pressure gradient which in turn will result in moisture vapour diffusion to the surface and this will frequently occur simultaneously with liquid moisture movement.

External Conditions: Direct-heated Dryers

Whereas an understanding of the internal mechanism of moisture flow during the drying of a solid is of importance in assessing the probable behaviour of a particular material, the mechanism of external evaporation at the solid surface and the effect of external variables are of major importance in investigating the drying characteristics of a material and in arriving at the correct type of dryer to employ and its operating conditions. The essential external variables are temperature, humidity, rate and direction of air flow, the physical form of the solid, the desirability of agitation and the method of supporting the wet solid during the drying operation. Some materials of a fibrous or porous nature will possess properties promoting rapid capillary and diffusion movement of moisture to the surface and the controlling factor will be the rate at which surface evaporation can be secured. In the initial stages of drying materials of high moisture content, also it is important to obtain the highest possible rate of surface evaporation. This surface evaporation is essentially the diffusion of vapour from the surface of the solid to the surrounding atmosphere through a relatively stationary film of air in contact with its surface. This air film, in addition to presenting a resistance to the vapour flow, is itself a heat insulant. The thickness of this film rapidly decreases with increase in the velocity of the air or gas in contact with it whilst never actually disappearing. The inner film of air in contact with the wet solid remains saturated with vapour so long as the solid surface has free moisture present. This results in a vapour pressure gradient through the film from the wetted solid surface to the outer air and, with large air movements, the rate of moisture diffusion through the air film will be considerable. The rate of diffusion, and hence evaporation, of the moisture will be directly proportional to the exposed area of the solid, inversely proportional to the film thickness and directly proportional to the difference in the vapour pressure of the moisture at the temperature of the inner film surface and the partial pressure of the water vapour in the surrounding air. It follows therefore that high surface evaporation rates will be secured through large exposed solid surface areas and high air velocities relative to the solid surface and low relative humidity of the drying air or gas. In practice, these conditions are produced through pre-forming, extruding or granulating the wet solid where appropriate; the use of fans to promote forced convection; and control of fresh air fed to the drying plant in quantities sufficient to attain optimum humidity conditions. As the heating up of a solid and the evaporation of the moisture entails the absorption of calculable and large quantities of heat, the latter must be supplied either from the air supply in the case of forced convection dryers or externally heated surfaces in the case of indirect dryers. It is of importance to note at this point that, since the layer of air film in contact with the wetted solid undergoing drying remains saturated at the temperature of the area of contact, the temperature of the solid surface whilst still possessing free moisture will lie very close to the wet-bulb temperature of the air.

The Periods of Drying

A necessary preliminary to the consideration of a drying problem, the selection of a suitable type of dryer and the sizing thereof, is the experimental determination of the drying characteristics of the wet solid. This is particularly so in the case of a forced convection dryer. A series of drying tests will be carried out under predetermined conditions which may be set on an empirical basis but will usually be influenced by what is already known as to the material's sensitivity to temperature and the limits of temperature attainable with the source of heat to be employed on the commercial units. In these tests, the effects of variable conditions are studied to ascertain optimum results and the rates of drying are determined under these conditions. The variables, as previously stated, will be temperature, air flow, material subdivision and air humidity. In order to reduce the number of variables it is usual before embarking on drying tests to set certain standard conditions common to a series of tests. Thus in the case of testing for drying characteristics under conditions of forced convection, it is usual to work with a common standard area of sample, all samples in a series being of uniform state of subdivision unless of sheet or cake form. The thickness of the bed of material on a supporting tray will also be maintained a constant unless the effect of differing bed thicknesses requires to be investigated as may often be the case. A series of tests will then be carried out to determine the rate of drying using different air temperatures and air velocities. It is assumed that the relative humidity of the ambient air remains constant during the tests. These tests will indicate the optimum combination of air temperature and flow to secure the desired dry product in the minimum of time. Determinations will be made of the moisture content of the sample W against the time of drying θ. If now the rate of moisture loss is plotted against time as shown in Fig. 2, a curve is obtained which indicates that the drying rate varies considerably during the course

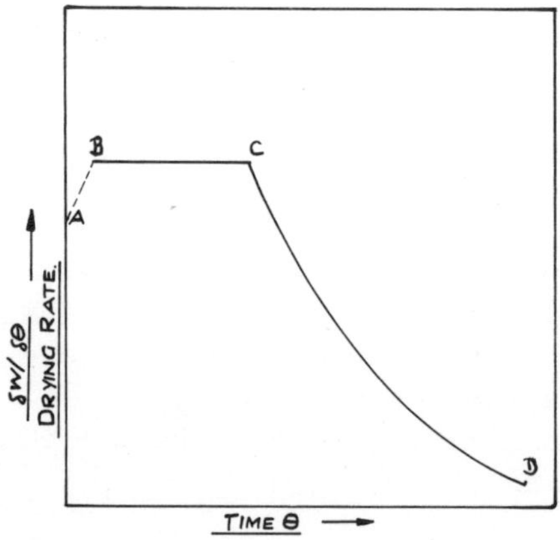

Fig. 2. The periods of drying.

of drying and is in no way constant. Most materials which are surface-wet will show three distinct stages in the drying-rate curve, namely, section AB where the wet material is warming up; section BC where the rate of drying remains substantially constant; and section CD where the rate of drying falls away in an ever decreasing degree. Section BC of the curve is referred to as the period of constant drying rate whilst section CD is referred to as the period of falling drying rate. Point C at which the constant rate ends and the falling rate commences is termed the point of critical moisture content. It is important to appreciate the significance of these two periods.

Constant Rate Period

During this period which exists for materials having free surface-wet moisture, evaporation is taking place from the surface of the solid and so long as this remains surface-wet is independent of internal mechanisms within the solid. The rate of drying is essentially that of the evaporation of the liquid component as though no solid were present under the conditions of temperature and air flow obtaining during the test. The temperature of the liquid film and that of the adjacent solid surface will remain substantially constant and will approximate to the wet-bulb temperature of the moving air at the test temperature assuming radiation and conduction effects are negligible. In practice, on an operating convection drying plant, these two effects are unlikely to be wholly absent and will result in a higher surface temperature than the wet-bulb temperature and will produce a higher constant drying rate. In the case of indirectly heated dryers, where heat is conducted to the wet solid by contact between the latter and the heated metal surface of the dryer, and where heat transfer by convection is insignificant, the temperature of the solid will gradually attain the boiling point of the liquid component and the drying rate during the constant rate period will be higher than that under solely convection drying conditions with the same air dry-bulb temperature as that of the metal surfaces. Higher drying rates in the constant rate period may therefore be anticipated with indirectly heated dryers provided the solid can be kept in a constant and intimate state of agitation in contact with the heated metal surface. Radiation, similarly, can result in higher solid temperatures and increased drying rates. In most commercial convection dryers, it can be assumed that metal surfaces in contact with the material being dried and those that have a "view" of the material will cause conduction and radiation effects and except in so far as the increased solid temperature produced is not deleterious to the product may constitute an additional factor of safety on the designed size of dryer. Summarising the above, it is evident that the rate of drying during the constant rate period depends on three externally operating factors, namely, the heat- or mass-transfer coefficient, the area of solid exposed to the drying medium, and the difference between the temperature and humidity of the drying medium and that of the wet solid surface. Increase in any or all of these factors results in increased drying rates.

Heat and Mass Transfer

The rate of heat transfer from the heating medium to the wet solid in a commercial dryer will depend upon the degree of agitation of the solid in relation to the heating medium and, in the case of forced convection dryers, the velocity of the air or gases used to apply heat. In indirect dryers, it is important to have the maximum contact between the heated dryer metal surfaces and the solid undergoing drying and the maximum relative movement at areas of contact. Indirect dryers, for most solids, will therefore be designed to give this effect through agitation, cascading or rolling action. In the case of forced convection or direct dryers, the principal factor affecting the rate of heat and mass transfer will be that of the hot air or gas velocity. High air velocities will reduce the thickness of the stationary gas film on the surface of the solid and hence increase the heat- and mass-transfer coefficients. In the practical designing of dryers it is found to be more reliable to consider heat-transfer rates than mass-transfer rates, as the latter are a function of the surface temperature of the wet solid which is difficult to determine and cannot, in practice, be assumed to be that of the wet-bulb temperature of the air with an adequate degree of accuracy, due to the effects of conduction and radiation. The heat-transfer coefficient in forced convection dryers where the hot air movement is parallel with the surface of the wet material, as in cross-flow tray dryers, can be expressed[1] in terms of mass velocity of the heating medium with a reasonable degree of accuracy; thus

$$h_c = 0{\cdot}0128\, G^{0{\cdot}8}$$

where h_c is the convection heat-transfer coefficient in Btu/h/ft^2/deg F and G is the mass velocity of the drying air in lb/h/ft^2. In the case of air blown perpendicularly on to the surface of the solid surface the heat-transfer coefficient is greater and has been shown[2] to be covered by the expression

$$h_c = 0{\cdot}37\, G^{0{\cdot}37}$$

Such conditions give an increase in the heat-transfer coefficient which is further enhanced at the very high mass velocities locally attainable through the use of slots or jets. Both the above expressions ignore the effects of conduction and radiation. Ignoring the heat required to raise the temperature of the bone-dry solid, which will be low relative to the total requirement in the case of wet solids having a constant rate period, a heat balance on the drying operation will yield the expression

$$\frac{\delta W}{\delta \theta} = \frac{h_c\,(t_a - t_w)}{d_s L \lambda}$$

where $\dfrac{\delta W}{\delta \theta}$ = drying rate, constant period, lb water/h/lb dry solid

t_a = dry-bulb temperature of drying air, °F

t_w = wet-bulb temperature of drying air, °F

d_s = bulk density of bone-dry solid, lb/ft^3

L = thickness of solid bed, ft

λ = total heat of vaporisation of water at t_w

The above expression assumes that evaporation is taking place from one surface of a layer of material supported on a solid tray. If evaporation is taking place from both surfaces as in the case of wire-mesh trays L = half the total bed thickness.

As has been mentioned earlier, in commercial forced convection dryers the effects of heat transfer by conduction and radiation may be appreciable, due to the fact that the material surface temperatures are higher than the wet-bulb temperature of the drying air. Thus in compartment or tunnel tray dryers, the drying rate in the constant rate period may be 20% to 40% higher than that calculated from the above expression through the effect of radiation. This fact emphasises the importance of carrying out pilot-scale tests on adequate samples of wet material with a full complement of loaded trays in the test unit. The above discussion also brings out the necessity for carrying out these tests not only at the maximum permissible air temperature but also at different air velocities and bed thicknesses in order to arrive at overall optimum drying conditions during the constant rate drying period.

In the case of through-circulation dryers, consideration of the drying rate during the constant rate period is more complicated as it is a function of several factors. Not only does it depend on the rate of air flow, air temperatures and humidities, but also the particle size and the pretreatment given to the wet material in order to produce a permeable bed. This pretreatment, whether granulating, extruding or pelletising, will determine the total exposed drying surface produced; extrusion may additionally cause considerable compacting of material within the extrusions. For these reasons, and in view of the fact that drying rates can be obtained easily and quickly in a suitably designed test unit, it is the normal practice to determine the drying characteristics for through-circulation dryers by adopting a test procedure. This involves passing heated air through a fixed area of material at different air temperatures and velocities and at different bed depths. Tests will quickly show what air velocity the bed of material will stand without entrainment of material when the effects of the other variables can be studied using the optimum air velocity. Through-circulation dryers are normally designed to operate at about 200–300 ft/min superficial air velocities over the effective area of the supporting perforated or wire mesh tray or conveyor.

Falling Rate Period

As has been stated, the period of constant rate of evaporation from a wet solid is followed by a period during which the rate of drying progressively decreases, the transition from one period to the other taking place at the point of critical moisture content of the solid. In practice, the original feedstock may have such an initial moisture content and the product may be required with such a high residual moisture content that all the drying will take place during the constant rate period. More usually both phenomena will be present and with very slow drying materials most of the drying operation may exist in the falling rate period. Most materials, as presented to a commercial dryer, will not show quite as sharp a change in the drying rate curve as that indicated by Fig. 2. Whereas during the constant rate drying period, if it exists, all the surfaces of the exposed solid are completely wetted, at the change to the falling rate period some of the solid surfaces will be still wet and some dry depending largely on the physical form of the material being dried. The rate of evaporation of the less moist surfaces will be lower than that of the completely wetted portions, the net result being a falling off in the rate of drying as drying proceeds when compared with the rate during the constant rate period. The rate of drying in this section of the drying curve will still be affected by factors which influence the constant rate drying period as discussed above. Once all the exposed surfaces of the solid cease to be wetted, however, the rate of drying will be a function of the rate at which moisture or moisture vapour can move physically by diffusion and capillarity from within the solid to its surface. Where considerable shrinkage occurs during drying, pressure gradients set up thereby within the solid will also assume importance.

Evidently, where the initial moisture content of a feedstock is relatively low and the final moisture content required is extremely low, the falling rate period becomes of predominating importance. Drying times will be long and the variables affecting drying rate in the constant rate period cease to be of such great consequence. Air velocities will be important only to the extent to which they enhance heat-transfer rates. Air temperature, humidity, material thickness, or bed depth become all-important. When the rate of diffusion is the controlling factor, particularly when long drying periods are required to attain low final moisture contents, the rate of drying during the falling rate period varies as the square of the material thickness, which indicates the desirability of granulating the feedstock or otherwise pre-forming it into small shapes, exposed in thin layers in the case of cross-flow tray dryers. Even so, extremely low final moisture contents for materials showing predominantly falling rate characteristics will usually demand considerable drying times, and hence higher capital investment in the installed dryer. It is important, therefore, at the outset, in considering a drying problem to establish what final moisture content is permissible and acceptable. Where a product attains equilibrium with the normally experienced relative humidity of the surrounding atmosphere in storage or in packages at a certain moisture content, the specifying of a final moisture content of the product discharged from the dryer below this figure is rarely justified in view of the extra cost of the dryer which would be required. Subsequent processing of the product – as for instance the performance of pulverising equipment – may of course dictate lower moisture contents.

Some materials with appreciable moisture contents, such as ceramic products, timber and soap, have drying characteristics over most of their drying curves such that the rate of moisture movement from within to the exposed surfaces is the controlling factor. As the rate of this moisture movement is a function of the temperature gradient from the solid interior to the surface, the higher the latter, the greater the drying rate. Further, where shrinkage is considerable or where the material may form a hard impermeable skin, it is desirable to retard the rate of evaporation to bring it into step with the rate of moisture movement to the surface. The material temperature at its surface will not be below the wet-bulb temperature of the drying air and will probably lie somewhere between this and the dry-bulb temperature due to radiation effects. The material surface temperature may be increased by increasing the wet-bulb temperature, i.e. reducing the wet-bulb depression of the drying air, through an increase in the humidity of the air. This, at the same time, has a controlling effect on the rate of surface evaporation due to the reduced humidity driving force promoting evaporation. Thus, for materials which have to retain their original shapes and dimensions, or to avoid surface cracking and splitting, the use of controlled humidity drying is to some degree desirable. Similarly in the drying of pastes and filter cakes, surface "case hardening" effects can be avoided through control of the humidity conditions of the air within the dryer.

Indirect-heated Dryers

In indirect-heated dryers, the drying operation does not depend upon the effect of convection from heated air or gases. Air flow through such dryers is merely that required to remove moisture vapour from the system as it is liberated in the course of drying. Heat transfer, and hence the drying operation, occurs as a result of heat conducted to the wet solid through the walls of the retaining vessel; the drying rate will therefore be a function of the heat-transfer rate by conduction, with the added effect of radiation where this occurs. This situation occurs with indirect rotary dryers, agitated pan and trough dryers, film drum dryers, vacuum dryers generally, steam tube rotary dryers and continuous sheet or fabric drum dryers. Radiation dryers – so-called infra-red dryers – are radiation dryers with or without a measure of forced convection: heat transfer by conduction is of no consequence in such designs.

Usually, with indirect dryers, through the action of heated contact surfaces, the temperature of the solid undergoing drying is higher than the surrounding air or neutral atmosphere and heat is more likely to be imparted to the air by the heated solid than the reverse. The material subjected to indirect drying will of course possess its normal drying characteristics in the form of its constant rate drying period and its falling rate period. External variables will, however, act quite differently from those which obtain in direct-heated dryers.

In the constant rate period, the rate of drying will be a function of the total heat-transfer surface in contact with the wet material; the temperature of the heating medium (usually steam or combustion gases); the temperature of the solid itself and the heat-transfer coefficients of the heating medium, the metal retaining wall, and the wall to wet solids. In considering an indirect dryer operation, these individual heat-transfer coefficients are unlikely to be known and would be difficult to assess. It is customary to consider the problem in terms of an overall heat transfer coefficient. On this basis, the total heat transfer (Q in Btu/h) would be expressed by

$$Q = kA\ (t_1 - t_2)$$

where k = overall heat-transfer coefficient Btu/h/ft^2/deg F temperature difference between the heating medium and the solid

A = total area of heat transfer between heated surface and wet material in ft^2

t_1 = temperature of heating medium (average), °F

t_2 = temperature of solid undergoing drying, °F

The value of k, the overall heat-transfer coefficient, will depend upon the type of dryer and its mechanical design, degree of agitation, or speed of revolution; adequate condensate drainage in the case of steam-heated dryers or combustion gas turbulence in the case of fuel-fired dryers; the temperature to which the retaining wall surface can be raised; and the physical properties of the feedstock. Normally experienced values for k will be indicated in the discussions on individual types of indirect dryers but are tabulated for convenience at this stage in Table 7. As the material will pass through the normal stages of constant and falling rate drying periods, its temperature will remain substantially constant during the former but will increase during the latter, so that in practice Q is evaluated on the basis of an average temperature difference between the wet material and that of the heating medium.

In jacketed or steam-tube heated indirect dryers, the temperature of the heating medium t_1 will remain constant and during the constant rate period the temperature difference between heating medium and solid $(t_1 - t_2)$ will remain substantially constant. In the case of indirect combustion gas-heated rotary dryers, the value of t_1 will vary along the length of the dryer. However, during the falling rate period the temperature of the solid will rise, so that in practice the value of k is usually assessed for the entire drying period as an overall heat-transfer coefficient based on the mean temperature difference between material and heating medium.

Equilibrium Moisture Content

It has been pointed out earlier that in considering a drying problem, it is important to establish at the earliest stage, the final or residual moisture content of the product which can be accepted. This is important as many

Table 7. Overall heat-transfer coefficients (k) for indirect dryers Btu/h/ft^2/deg F average temperature difference

Type of dryer	k
Vacuum shelf dryer	1
Agitated pan dryer	5–60
Rotary vacuum dryer	5–50
Indirect rotary dryer	2–10
Jacketed trough dryer	2–15
Drum dryer	200–300

materials are to some extent hygroscopic and if dried below a certain moisture content they will absorb or "regain" moisture from the surrounding atmosphere depending upon its temperature and humidity. The material will establish a condition in equilibrium with this atmosphere and the moisture content of the material under this condition is termed the Equilibrium Moisture Content. Equilibrium moisture content is not greatly affected at the lower end of the atmospheric scale but as this temperature increases the equilibrium moisture content figure decreases, which explains why materials can in fact be dried in the presence of superheated moisture vapour. Drying medium temperatures and humidities assume considerable importance in the operation of direct dryers, e.g. during the drying of certain inorganic salts to a specified content of water of crystallisation, as does the particular design of dryer employed: continuous parallel flow pneumatic and rotary dryers facilitate control of the temperature and humidity of the atmosphere in contact with the product during the course of drying and hence the equilibrium conditions between product and drying medium.

In the more common drying operations met with in practice, the equilibrium moisture content of a material is important as drying may be carried out unnecessarily far, resulting in a reduction in the capacity of a given drying installation and an unjustifiably high cost of drying. Thus, if the equilibrium moisture content of wool hanks in contact with normal ambient air is of the order of 13 to 14% on the wet-weight basis, there would be no point in drying to a much lower moisture content, thus sacrificing the capacity of the dryer to dry more wool in a given time and consuming more steam in the heater batteries than is necessary. Table 8 gives the equilibrium moisture content of a number of materials in contact with air at different percentage relative humidities at 60°F.

Psychrometry in Relation to Drying

As has been stated previously, two processes occur simultaneously during the thermal process of drying a wet solid, namely, heat transfer in order

Table 8. Equilibrium moisture content of some materials (percentage on dry-weight basis)

Material	*Percentage relative humidity of air*				
	20	30	40	50	60
Wool (worsted)	9·5	13·0	16·5	18·5	21·5
Cotton cloth	3·5	4·5	6·0	7·0	7·5
Egyptian cotton	3·5	4·5	5·5	6·0	7·25
Linen	2·75	3·5	4·5	5·1	6·0
Press board	3·5	4·5	5·5	6·25	9·5
Sole leather	11·5	13·5	15·0	16·0	17·5
Flour	3·95	5·0	7·0	8·5	10·0
Biscuit	2·5	3·0	4·0	5·2	6·5
Ball clay	2·5	3·0	3·5	4·0	4·3
Kaolin	0·7	0·8	0·9	0·95	1·0
Asbestos	0·7	0·8	0·9	0·95	1·0
Zinc oxide	0·35	0·4	0·4	0·4	0·45
Soap	4·0	5·0	6·5	8·0	10·5
Glue	5·0	6·0	6·75	7·5	8·6
Wood	4·5	6·0	7·25	9·0	11·25

to raise the temperature of the wet solid and to evaporate its moisture content together with mass transfer of moisture to the surface of the solid and its evaporation from the surface to the surrounding atmosphere which, in convection or direct dryers, is the drying medium, usually heated air or combustion gases. The questions of heat transfer and mass transfer through the solid have been considered and the effect of external variables has been discussed. Consideration of the actual quantities of air required to remove the moisture liberated by evaporation is based on psychrometry and the use of humidity charts. This procedure attains considerable importance in the case of forced convection, pneumatic and direct rotary dryers. In the designing of direct-heated dryers the estimates based on heat requirements are easily arrived at as they depend upon the quantities of bone-dry solid and moisture involved, specific and latent heats and operating temperatures. The quantity of air required to remove the moisture liberated, as distinct from the quantity of air which will release the required amount of heat through a drop in its temperature in the course of drying, however, has to be determined from the known capacity of air to pick up moisture at a given temperature in relation to its initial content of moisture. For most practical purposes, moisture is in the form of water vapour but the same principles apply, with different values and humidity charts, for other volatile components. Further, for most practical purposes, humidity charts are drawn on the assumption that the pressure of air is at or near atmospheric.

Air at a given temperature will hold a known maximum amount of moisture vapour when it is saturated with that vapour. It is usual to consider humidities in terms of weight rather than volumes. The following are definitions of expressions used in psychrometry.

ABSOLUTE HUMIDITY (H) is the number of pounds of moisture carried by one pound of bone-dry air. This is at a maximum when the air is saturated with moisture at a specified temperature and is represented on a humidity chart as the saturation curve obtained by plotting the humidity of the air against the dry-bulb temperature of the air. This is the maximum amount of moisture which the air will carry at the specified temperature when the partial pressure of water vapour in the air is equal to the vapour pressure of water at that temperature. In terms of weight units this, by definition, is recognised by

$$H_s = \frac{p_s}{P-p_s}\left(\frac{18}{28{\cdot}9}\right)$$

and the absolute humidity at any condition less than saturation will be:

$$H = \frac{p}{P-p}\left(\frac{18}{28{\cdot}9}\right)$$

where H_s = saturation humidity in lb/lb dry air
H = humidity lb/lb dry air
P = absolute pressure
p_s = vapour pressure of water at temperature t_s
p = partial pressure of water vapour in air at temperature t
$\frac{18}{28{\cdot}9}$ = ratio of molecular weight of water to air

Percentage Absolute Humidity may be written as

$$H_A = \frac{p}{p_s}\left(\frac{P-p_s}{P-p}\right)100$$

and Percentage Relative Humidity will be

$$H_R = \frac{100p}{p_s}$$

by eliminating the factor

$$\frac{P-p_s}{P-p}.$$

In humidity charts Absolute Humidity values are plotted against dry-bulb temperatures and it follows that a series of curves will be obtained for

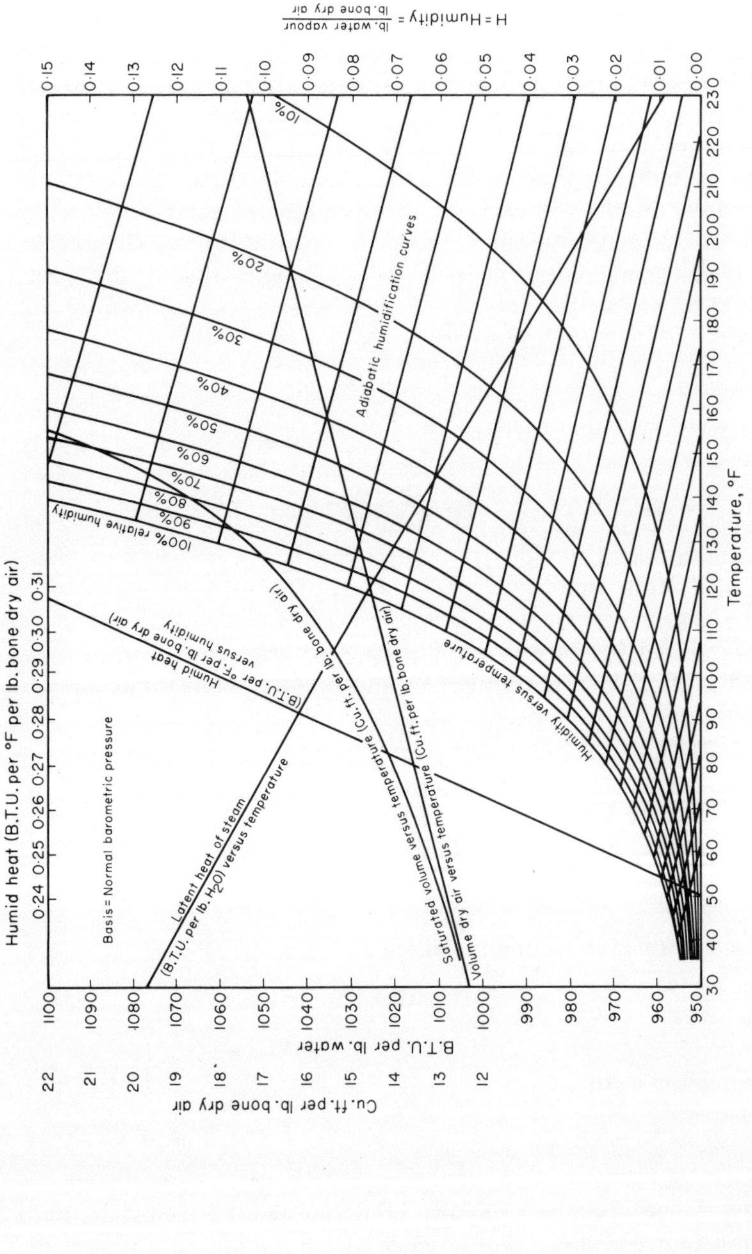

Fig. 3. Humidity chart.

Courtesy McGraw-Hill Book Company Inc.

different percentage humidity values from saturation downwards. Fig. 3 illustrates a typical humidity or psychrometric chart showing certain additional data.

DEW-POINT is the temperature of the air at which it attains a condition of saturation; the dew point for any air temperature will therefore fall on the saturation curve.

ADIABATIC COOLING LINES. An adiabatic process is one in which heat interchange with the surroundings is negligible. Most industrial dryers will be well insulated in order to reduce heat losses to a minimum and, for purposes of using humidity charts in dryer calculations, it is assumed that adiabatic conditions exist, correction being added later to cover expected heat losses. Adiabatic cooling lines are shown on a humidity chart as straight lines starting from the 100% saturation curve and running down to the right. When air of a given humidity cools adiabatically in contact with water at the wet-bulb temperature of the air its humidity increases; the fall in temperature and the rise in humidity follows the path of the adiabatic cooling lines terminating at the saturation curve. There will be a series of such lines, parallel to one another, for differing initial humidities. When unsaturated air is passed over water which is at a lower temperature than the air, the air temperature will fall and, under adiabatic conditions, the heat which it gives up in cooling is transmitted to the water, raising it to the wet-bulb temperature and, if the heat is sufficient, water is evaporated thus raising the humidity of the air. If the water is initially at the air wet-bulb temperature, the heat balance on the process, assuming one pound of dry air, will be represented by the expression

$$\lambda \delta H = -c_s \delta t$$

where δH = differential weight of water evaporated in pounds
δt = differential drop in temperature of the air in °F
c_s = humid heat of the dry air in Btu/lb/deg F
λ = latent heat of evaporation of water at the wet-bulb temperature in Btu/lb

The minus sign indicates that t falls with increase in H.

Rewritten as $\delta H/\delta t = -c_s/\lambda$ it is seen that the slope of the humidity versus air temperature curve is equal to the ratio c_s/λ; the minus sign indicating its slope is downwards to the right from the horizontal. For relatively small increases in humidity the values of c_s and λ are substantially constant, so that the curve of H plotted against t, i.e. the adiabatic cooling curve, is substantially a straight line the slope of which is $-c_s/\lambda$. The humid heat of air c_s is the sum of the heat content of bone-dry air and its water vapour content at a given temperature and is given by

$$c_s = 0{\cdot}24 + 0{\cdot}466H$$

where c_s = humid heat Btu/lb dry air/deg F
0·24 = heat capacity of dry air Btu/lb/deg F
0·466 = heat capacity of water vapour Btu/lb/deg F
H = absolute humidity of air lb/lb dry air

Humid heat is represented by a straight line starting at 0·24 for zero humidity and running upwards towards the saturation curve.

The adiabatic cooling curves on a psychrometric chart assume ideal adiabatic conditions and that the latent heat of evaporation is exactly provided by the temperature drop of the air streams, the wet-bulb temperature of the air stream remaining constant. In practice, of course, heat losses occur and heat is required to raise the temperature of the wet material to its discharge temperature. The drying operation is not therefore in practice truly adiabatic and a suitable correction has to be applied to take care of these additional heat requirements in assessing the amount of air required. Instead of the air moving along a simple ideal wet-bulb line as shown on an ordinary psychrometric chart, it moves along a line at a flatter angle than that corresponding to $-c_s/\lambda$ and for the same exit temperature terminates at a lower wet-bulb temperature and contains less moisture. The angle between the ideal wet-bulb line and that actually followed by the air-stream conditions can be shown to be a fairly simple trigonometric function of the additional heat requirements divided by the weight of water evaporated, and it is sometimes the practice to construct a special psychrometric chart from which the actual air requirements can be calculated direct from convection drying tests on pilot-scale units based on inlet and outlet air conditions, the weight of water evaporated and the calculated heat losses from the pilot-scale unit and that of the anticipated size of commercial sized plant for the required duty under test unit conditions (see Fig. 5).

SPECIFIC VOLUME. Psychrometric charts usually indicate two curves giving the relationship between the volume of a given weight of air and its temperature, known as the specific volume. One of these—a straight line curve—gives the specific volume of bone-dry air whilst the second gives the specific volume of saturated air. The line for bone-dry air is given by the expression

$$V_d = 11{\cdot}57 + 0{\cdot}025\,t$$

whilst that for saturated air is given by

$$V_s = 0{\cdot}0405(460+t)(0{\cdot}622+H_s)$$

where V_d and V_s = specific volume of dry and saturated air, respectively, in ft^3/lb dry air
t = temperature of air, °F
H_s = saturation humidity lb/lb dry air.

The specific volume of air at a given percentage relative humidity and temperature is obtained by reading off the dry air and saturated air specific volumes from the two curves at the specific temperature and adding together the products of these two figures in proportion to the components – bone-dry air and water vapour – as given by the percentage relative humidity of the air.

CHAPTER III

Types and Classifications of Industrial Dryers

A wide variety of dryer designs have been evolved over the years with a view to carrying out the unit operation of drying. Several methods have been suggested as to how best to classify them but before considering seriously the best method of classification it is necessary to define on broad principles how the necessary thermal requirements are supplied and the types of industrial dryers which are today available.

In the first place, heat must be transferred to the wet material to promote the drying operation. Heat can be applied by one or more of the following methods:

1. *Convection*, whereby the heating medium, usually air or the products of combustion, is in direct contact with the wet material.
2. *Conduction*, whereby heat is transmitted indirectly by contact of the wet material with a heated surface.
3. *Radiation*, whereby heat is transmitted directly and solely from a heated body to the wet material by radiation of heat.

Heat supplied by radiation alone in a drying operation is on the whole exceptional and is usually restricted to so-called infra-red dryer installations which will be described in due course. Radiation effects in drying operations are usually incidental, and sometimes accidental, but may have quite an important bearing on the performance of a dryer in which convection or conduction or both effects are designed to predominate. The majority of industrial drying installations rely on the heat transfer and drying effect through convected heat supplied directly to the wet material by a mass of heated air or combustion gases forced or circulated in contact therewith by means of fans or by the conduction of heat to the material through a retaining enclosure heated externally from a source of heat, water vapour released in the drying operation being removed by a mild exhaustion of humid air.

Broadly speaking, therefore, industrial dryers fall into two main categories, namely, *convection* dryers and *conduction* dryers, with additional special types involving almost entirely radiation or some other specialised form of heat generation or application.

Such dryers will also take the form of intermittent or "batch" type dryers or, alternatively, continuous dryers and we have therefore two categories of convection dryer and two categories of conduction dryer. In order to familiarise the reader at this stage with the wide variety of industrial drying plant falling within these categories, the following is a brief description of each with illustrations of typical units. A more detailed treatment of each type and variants thereof is given later in the book.

Convection Dryers

(*i*) *Batch*

The simplest and hitherto most widely used industrial dryers have been drying stoves, ovens or cabinets into which trays or pans of wet material are charged, either directly or supported on mobile trucks, and are allowed to remain in the dryer under conditions of natural or forced convection (through the use of fans) until the drying operation is complete, Plate I. The drying and heating medium is usually air which has been heated by a steam-heated exchanger, the path of air being transversely across the areas of the trays, or in special cases through the material supported on perforated or wire-mesh trays. More recently, in order to reduce both handling costs and drying times, the wet charge is supported in a receptacle having a perforated or permeable base through which sufficient hot air at adequate pressure is forced in order to fluidise the charge. Both these types are batch convection dryers.

(*ii*) *Continuous*

The majority of accepted dryer designs fall into this category and include the following.

Truck-and-Tray Tunnels, Plate II, frequently of considerable dimensions in which a number of laden trucks are propelled from the charging to the discharging end under forced convection conditions similar to those obtaining in batch tray dryers. Such tunnels are frequently designed to operate quite automatically in fully continuous or semi-continuous manner, the trucks and their charge being propelled through a temperature and time cycle appropriate to the drying characteristics of the charge. A more truly continuous tray dryer is the *rotary tray dryer*, Plate III, which is composed

Plate I. Batch tray dryer. *Courtesy A.P.V.-Mitchell (Dryers) Ltd.*

Plate II. Continuous truck and tray tunnel dryer. *Courtesy A.P.V.-Mitchell (Dryers) Ltd.*

of several tiers of rotating trays, the wet feedstock being fed on to the top tier and the material being raked through a radial aperture on each tray so that it descends from tray to tray to its final discharge point. A central shaft generally carries a number of vaned impellers which circulate air over the trays through heating coils at the periphery of the trays.

Continuous Through-circulation or *Band Dryers* are composed of either a single or several perforated or wire-mesh metal aprons or conveyor bands on to which the wet charge is evenly distributed at the feed end whence it passes through the length of the dryer, the heated drying air being forced through the bed of material. Dried product is discharged continuously at

Plate III. Continuous rotary tray dryer. *Courtesy Buell Ltd.*

the exit from the dryer. The temperature and humidity of the drying medium (air) can be adjusted throughout the passage of the material through the dryer to satisfy its drying characteristics.

Rotary Dryers, Plate IV, are essentially slowly rotating cylindrical shells provided with means for continuously feeding the wet charge, propelling it through the dryer with a lifting and showering effect and finally discharging the dried material from the dryer in a continuous manner. The drying and heating medium, heated air or combustion gases, is fed into the dryer shell where it comes into contact with the showered charge, the movement of the gases being either co-current or counter-current with the movement of the charge.

Fluid Bed Dryers, Plate V, operate on the same principle as batch type fluidised bed dryers except that the flow of wet material into the dryer and dried product out of the dryer is continuous, the charge in the fluidised form having the characteristics of a fluid. Such dryers usually take the form of vertical cylindrical columns but rectangular designs are sometimes used.

Spray Dryers, Plate VI, are used for drying liquid solutions, suspensions and pumpable pastes by bringing them into intimate contact in the form of

Plate IV. Continuous rotary dryer. *Courtesy A.P.V.-Mitchell (Dryers) Ltd.*

Plate V. Continuous fluid bed dryer. *Courtesy A.P.V.-Mitchell (Dryers) Ltd.*

Plate VI. Spray Dryer. *Courtesy A.P.V.-Mitchell (Dryers) Ltd.*

small droplets with heated air or combustion gases in a suitably designed enclosure. They are more particularly suitable for large productions of a single product and where the material is relatively heat-sensitive. Evaporation of the moisture in droplet form is almost instantaneous resulting in rapid reduction of the gas temperature with moderate increase in material temperature. The dried product is almost always in spherical form, the size of which is amenable to a certain degree of control. The relative movement of feedstock and drying medium can be co-current, counter-current, or a combination of both.

Pneumatic or Flash Dryers, Plate VII, rely on total entrainment of the feedstock presented in a suitable form in a current of heated air or combustion gases and the pneumatic conveyance of the material in the process of drying co-currently with the drying medium to a suitably designed solids-gaseous separation system such as a cyclone in which drying is or may be completed. A measure of both gas and material recycling may be adopted.

Plate VII. Pneumatic or flash dryer. *Courtesy Rosin Engineering Ltd.*

Conduction Dryers

(*i*) *Batch*

Vacuum Tray Dryers consist of an insulated and sometimes heated enclosure into which trays or pans of wet material are introduced and supported on steam-heated platens or electrically-heated plates. The enclosure is provided with a quick-acting vacuum proof door together with dust catchpot, condenser, and vacuum pump. Heating of the wet material is by conduction of heat from the supporting platens. Freeze dryers are a special form of this type of dryer.

Vacuum Double Cone Dryers, Plate VIII, are rotated about their axis and a charge of wet material is introduced into them. They are generally steam-jacketed and may be provided with internal mechanical means for breaking up the charge during the process of drying. They may be designed for atmospheric or reduced pressure operation.

Trough Dryers usually take the form of jacketed U-troughs provided with mechanically driven paddles or spiral ribbons. They may be operated at either atmospheric or reduced pressure.

Plate VIII. Vacuum double cone dryer. *Courtesy Patterson Division, Banner Industries, Inc.*

Pan Dryers, Plate IX, consist of shallow cylindrical vessels provided with means for agitating the charge. They can be used at atmospheric or reduced pressure and may be steam-jacketed, electrically heated, direct-fired or may have cast-in high pressure heating coils in the sides and base. They are usually arranged for top loading and may have either bottom or side discharge doors. The agitator may be either top or bottom driven through suitable gearing.

Rotary Dryers for vacuum operation, Plate X, are composed of a jacketed horizontal cylinder which is stationary, but provided with an internal shaft fitted with paddles or scraper blades. Both the shaft and agitator gear may be designed for steam heating. Depending upon the size of unit there will be one or more charging points above and usually two or more charging doors below.

(*ii*) *Continuous*

Film Drum or *Roller Dryers,* Plate XI, may be designed to operate under atmospheric or vacuum conditions and consist of either one or a pair of

Plate IX. Agitated pan dryer. *Courtesy A.P.V.-Mitchell (Dryers) Ltd.*

Plate X. Rotary vacuum dryer. *Courtesy A.P.V.-Mitchell (Dryers) Ltd.*

Plate XI. Film drum dryer. *Courtesy A.P.V.-Mitchell (Dryers) Ltd.*

machined and polished steam-heated cylinders rotating about their axis. By suitable feeding arrangements a film of liquid containing a solid in solution or suspension is formed on the drum surface on which it dries during the rotation of the drum, the dried material being removed by scraper or "doctor" knives.

Cylinder or "*Can*" *Dryers* operate on much the same principle as film drum dryers but consist of a battery of steam-heated drums over the surface of which paper or fabric is continuously fed, drying being the result of direct contact of the material with the steam-heated drum surface.

Rotary Dryers, Plate XII. Where drying in a continuous rotary dryer is as a result of conduction of heat through the dryer shell to the material being dried, this type is referred to as an "indirect" rotary dryer. It is usually heated by combustion gases passing around the external surface of the dryer shell, which is rotated and through which the material passes continuously as in the case of convection or "direct" rotary dryers. Lifting flights are provided but these will be much shallower than in direct rotary dryers as the use of this type of dryer is usually limited to fine, dust producing materials. In some cases internal longitudinal steam-heated tubes are used in place of combustion gases.

Trough Dryers are either of U-section or tubular, either steam-jacketed or for heating in a direct-fired flue in which the charge is continuously propelled through the dryer shell by means of agitator blades or a screw. In

Plate XII. Continuous steam tube rotary dryer. *Courtesy A.P.V.-Mitchell (Dryers) Ltd.*

some designs, the latter may be arranged to be steam-heated.

It has been suggested that there is a wide variety of methods of classifying industrial dryers and, indeed, reference to the literature will show that this is possible. However, from the practical point of view, a distinction should be made between classification and selection. The ultimate objective, after consideration of all aspects of the problem, is to select from a wide range of types available the most suitable type of dryer to accomplish the drying operation. These types are best analysed and classified according to their method of operation, leaving to selection their suitability for specific materials and under stipulated conditions. On this basis, the most appropriate method of classification will be related to the method of heat transfer and to whether the dryer operates in an intermittent or continuous manner. Conduction dryers will be further subdivided into atmospheric and vacuum types. Thus:

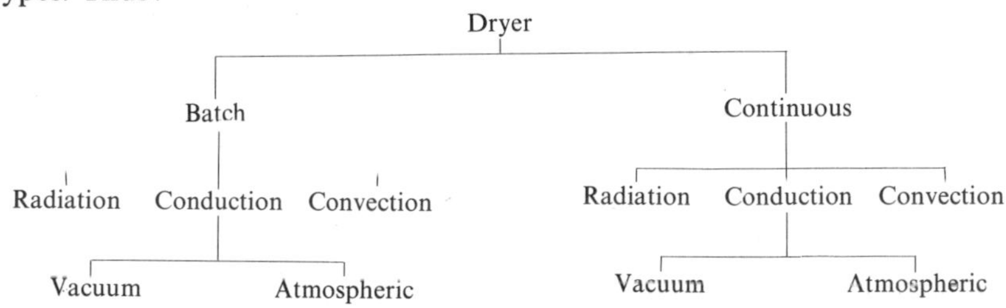

Each type of dryer generally available on the market today will now fall into place under this basic classification and the complete picture will take the form shown in Table 9.

Dryers which depend on heat transfer and drying by convection are usually referred to as direct dryers, and those depending on conduction as indirect dryers. Direct dryers usually rely on heated air or combustion gases but in some cases may be hot neutral gases or superheated vapours. In the case of combustion gases, their temperature may range up to about 1,500°F except where special heat-resisting alloys are used in the construction of the dryer when the temperatures may be higher. Obviously, at these high temperatures both radiation and conduction from metal and other surfaces will have an important bearing on the drying rate. Also at gas heating-medium temperatures appreciably above the boiling point of the moisture to be removed, the humidity of the heating and drying medium becomes less significant, hence the possibility of drying actually in contact with superheated vapour of the same composition as that of the liquid component removed in the drying operation. Conversely, at lower temperatures, humidity of the drying medium becomes important and in drying at low

Table 9. Classification of dryers,

Dryer			
Batch	Radiation		Infra-red
Batch	Conduction	Vacuum	Tray; Trough; Agitated pan; Double cone; Rotary
Batch	Conduction	Atmospheric	Tray; Trough; Agitated pan; Double cone; Rotary
Batch	Convection		Tray; Fluid bed
Continuous	Radiation		Infra-red
Continuous	Conduction	Vacuum	Band; Film drum
Continuous	Conduction	Atmospheric	Rotary; Film drum; Trough; "Can"
Continuous	Convection		Rotary; Band; Tray-tunnels; Fluid bed; Pneumatic; Spray; Rotary-tray; Conveyorised

temperatures, as for gelatines and glues, it is often desirable to dehumidify the drying medium – in this case, air.

Conduction dryers are usually referred to as indirect dryers and the heating medium may be condensing steam, electrical heating elements, combustion gases, hot liquid heating media such as high pressure water or organic salts and liquids. The type of heating medium employed will depend on convenience, size of unit or the temperature desirable for the drying operation and may range from below freezing point (in freeze drying) to above 1,000°F in the case of indirect-heated rotary dryers. All batch vacuum dryers are indirect dryers and, with suitably designed feeding and discharging arrangements, continuous vacuum dryers will be indirectly heated. Vacuum dryers would normally be used where the liquid removed during drying is a valuable recoverable component, where oxidation has to be avoided, or where there is serious risk of explosion in the presence of air. The presence of a toxic component may also indicate the use of a vacuum indirect dryer. Very dusty materials are usually dried in indirect dryers. Wherever possible, a high degree of material agitation or disturbance should be aimed at in indirect dryers although in vacuum shelf dryers this will not be possible.

CHAPTER IV

Selection of Industrial Dryers

When faced with a drying problem on an industrial scale, many factors have to be taken into account in selecting the most suitable type of dryer to install and the problem requires to be analysed from several standpoints. Even an initial analysis of the possibilities must be backed up by pilot-scale tests unless previous experience has indicated the type most likely to be suitable. The accent today, due to high labour costs, is on continuously operating unit equipment, to what extent possible automatically controlled. However, such considerations as flexibility, market fluctuations, specification changes, tonnages and quality of labour available must be taken into account. In any event, the selection of a suitable dryer should be made in two stages, a preliminary selection based on the general nature of the problem and the material to be handled, followed by a final selection based on pilot-scale tests or previous experience combined with economic considerations.

Preliminary Selection

The first consideration is dependent on whether the material to be dried is solid, liquid or gaseous. If liquid or gaseous, the choice is at once limited. For the former, almost without exception, the dryer will be either a film drum dryer or a spray dryer, whereas if the product is gaseous either an adsorption, absorption or chemical desiccant dryer will be used. Furthermore, unless the quantities involved are extremely small the latter classes would employ continuously operating dryers. In the case of wet solids, the first consideration will almost invariably be that of quantity to be processed. If the production rate is relatively small, say, up to 100 lb/h, or if several products have to be dried in the same drying unit, the selected dryer will operate on the batch or intermittent principle. Large productions of the order of one ton or more per hour will indicate continuously operated dryers and this will hold equally where the production of dried product from a very high moisture content feedstock is relatively small. Between these two categories lies a somewhat indeterminate range of medium production rates where the advantages or disadvantages of batch versus continuous dryers must be considered.

Pharmaceuticals, dyestuffs, pigments and fine chemicals are usually manufactured in a multiplicity of types in the same works, often amounting to considerable tonnages, with the natural desire of the manufacturer to consider continuously operating dryers of adequate flexibility in performance on the grounds of economy of capital, cost and floor space, not to mention labour. The danger of cross-contamination in such cases is serious and will most usually result in a decision to install several batch operating dryers. A colour works manufacturing a range of pigments may deal in considerable tonnages spread over a range of products but the danger of cross-contamination, the intermittent nature of prior processes, the peculiar drying conditions demanded by one or more of the products and the vagaries of market demand may indicate a deciding preference for intermittently operated individual dryers for each product. Only when there is a considerable demand for one or two products at such a works would continuous dryers be justified. Where the product is in filter-cake or pulp form, it should also be remembered that a batch truck-and-tray drying oven has a very considerable packing density in terms of material at relatively low capital and maintenance cost for the dryer. As in such cases the drying cycle is usually protracted, once the dryer is charged it can be left to complete the drying cycle under automatic thermostatic control without attention, the labour involved in the charging and discharging operations being free to carry out other work; the thermal efficiency of the dryer, if correctly operated, is not seriously inferior to a continuous dryer. With materials of a crystalline or free-flowing nature having a reasonably low moisture content, shorter drying times can be obtained and the inconvenience of filling and emptying a large number of trays can be avoided by using a batch-type fluid bed dryer.

For medium production rates, especially when batches of material come forward for drying from intermittent centrifuges or filters, agitated pan dryers are often extremely suitable. In certain cases, the charge from a reactor can be dropped direct into such a dryer very conveniently even in a slurry or semi-liquid form. It would be necessary to match the dryer capacity with the production rate from the reactor so as to avoid any bottlenecks at the drying stage. Where it is expected or known that drying will have to be carried out under vacuum it can be safely assumed that the dryer will have to operate intermittently and it will usually take the form of an agitated pan dryer, a vacuum trough dryer, a vacuum rotary dryer or a vacuum double-cone dryer.

For large-scale productions, continuously operated dryers become almost imperative but it should be remembered that wherever possible these should be run without interruption for days, if not months, at a time. Continuous operation over long periods leads to higher thermal efficiency, lower operating costs, highest returns on capital and greater uniformity of product quality. Pneumatic, spray and film drum dryers do not suffer unduly from loss of thermal efficiency when operated over shorter periods but quality control is likely to suffer. They are, however, easy to start up and they settle down to steady operating conditions fairly quickly. Continuous dryers rely for their optimum performance on a continuous and uniform feeding of the wet feedstock to the dryer so that if the manufacturing process immediately ahead of the dryer is an intermittent one, as it often is, careful consideration has to be given to the method of converting an intermittent availability to a continuous feed to the dryer. This will usually involve a buffer storage capacity either in hoppers or on conveying equipment. Process and site considerations will therefore have a bearing on the choice of the most suitable continuous dryer.

Summarising the above, the first stage of preliminary selection would result in the following reasoning:

	Production	*Dryer*
Quantity	Small	Batch
	Medium	Batch or continuous
	Large	Continuous
Previous processing	Intermittent	Usually batch
	Continuous	Continuous
Subsequent processing	Intermittent	Batch
	Continuous	Continuous

Whether a batch or continuous dryer results from this stage of selection, the next important factor in selecting the most suitable design is the physical and chemical properties of the material to be handled, its drying characteristics, the desired qualities of the dried product and any special individual problems such as material losses, explosion or fire hazards and solvent recovery. In addition, the facilities available at the point of installation of the drying unit will have to be taken into consideration, such as space available, most convenient heat source, risks of contamination of product and effluent problems. Feedstock in the form of slurries, pastes, pulps, or thixotropic filter cakes will call for a different type of dryer from one which is suitable for crystalline or free-flowing materials. A material which easily suffers attrition in the course of drying, leading to dust problems, may call for a different type of dryer from one which is suitable for coarse products. The presence of toxic, corrosive or inflammable components will require special treatment. The drying characteristics and the final desired moisture content may demand long residence times within the dryer; heat sensitivity may demand short drying times at relatively low temperatures or longer drying times at reduced temperatures and under reduced pressure. It will be appreciated therefore that to arrive at a short list of possible dryers from the wide range available, several factors must be carefully considered, a summary of which is as follows:

1. Properties of the material both as fed to, during its behaviour within and as delivered from the dryer, together with the possibility of toxicity, inflammability and abrasion effects.
2. Drying characteristics, including initial moisture content relative to the critical moisture content, type of moisture – whether bound, unbound or both, permissible maximum material temperature, and the anticipated drying time.
3. Specification of the dried product, including its purity, retention of physical and chemical properties, final moisture content, size analysis and its bulk density.
4. Product losses both as dust and, where solvents are present, solvent losses.
5. Site considerations, including space and height available, energy availability, legislative restrictions on effluent, and atmospheric conditions.

Batch Drying of Solids

If, after preliminary consideration, a batch-type of dryer has been decided upon, the selection of the most suitable dryer is relatively straightforward. Where the material is not reasonably free-flowing and cross-contamination between a variety of products must be avoided, a tray drying oven or cabinet will be used. Laboratory drying tests will show whether drying can be carried out at atmospheric pressure, and only if the material is excessively heat-sensitive or for other very special reasons, such as solvent recovery or requisite physical characteristics in the final product, will a vacuum tray dryer be used.

For reasons of thermal efficiency a forced convection recirculation oven will be the choice unless there is a danger of explosion, fire or excessive corrosive effects, in which case a total rejection oven will be used. If the wet material is such that laboratory tests prove it capable of fluidisation, a batch fluid bed dryer is likely to be the choice due to shorter drying times and less manhandling of the product being possible. This would be particularly applicable in the case of pharmaceutical products, crystals, and fine chemicals. If, on the other hand, cross-contamination and absolute cleanliness of the equipment were not all-important, an agitated pan, double cone, trough, or rotary dryer might be selected to operate under either atmospheric or vacuum conditions as necessary. Where it is necessary to dry the material in a neutral atmosphere or where the moisture consists of an explosive or valuable volatile material, one of the latter types of dryer working under reduced pressure will be employed. If the material is fibrous or amenable to through-air circulation, but not suitable for fluidising, a through-circulation tray dryer may be used. Forced convection batch ovens would also be used for small products of sheets or boards, electrical and other components, textile hanks, skeins, packaged goods, ceramics, open-work paint sprayed articles. Batch infra-red ovens could be used for paint sprayed articles having larger areas. Table 10 (pages 58/59) indicates the probable selection of batch dryers based on the properties and nature of the materials handled.

Continuous Drying of Solids

Should the first stage of preliminary selection have indicated that a continuous dryer is justified it is next necessary to narrow the choice from a consideration of the physical characteristics of the wet feedstock. True liquids, such as solutions, thin suspensions, and slurries will indicate film drum dryers or spray dryers. Certain liquid mixtures and gases will call for adsorption, absorption, or chemical desiccant dryers. These cases will be dealt with later. Wet solids, from thick slurries to surface wet crystals or moist powders and granules, will usually call for one of the several designs of dryer specially designed to handle solids. This may entail back-mixing of some dry product or it may require physical pre-forming of the wet feedstock. The main consideration is that the material can be handled mechanically in a continuous manner into and through the dryer and that it can be brought effectively into intimate contact with the drying medium – heated gases in the case of direct dryers, heated surfaces in the case of indirect dryers. A pumpable slurry, paste, or sludge will require the maximum possible immediate moisture evaporation on its introduction into the dryer in order that it may break down into manageable particles of semi-dried material and not build up on internal surfaces of the dryer. Hence a direct fired parallel flow type of rotary dryer is the most likely choice. Contact between the hot combustion gases and the cold wet feed will result in almost instantaneous flash evaporation with limited rise in the material temperature but considerable reduction in the combustion gas temperature. Large temperature differences between ingoing and outgoing gas temperatures will result, leading to high thermal efficiencies. Medium pastes, hard pastes, and filter cake, whether with or without dry product back-mixing, may lend themselves to pre-forming mechanically into a physical form suitable for rotary, pneumatic, fluid bed, or continuous band drying. Such materials, if tests show them difficult to dry, requiring lengthy residence time, will call for a rotary, a rotary-tray, or a band dryer. Materials with easier drying characteristics may be satisfactorily handled in either a pneumatic or fluid bed dryer. Residence times in pneumatic dryers are measured in seconds whereas a fluid bed dryer has an appreciably longer residence time. However, materials which do not demand a lengthy residence time but which cannot be fluidised would be dried in a pneumatic dryer, even if this had to be of a multi-stage design in order to give the required residence time for the desired degree of drying.

Consideration of site conditions might determine whether the choice should lie between a rotary or band dryer on the one hand or a pneumatic or fluid bed dryer on the other. Lack of floor space but availability of height would weigh in favour of the latter two in preference to the former.

The physical nature of the material during the course of drying will have an important bearing on the choice of type of dryer. Should the material break down to extremely fine particle size by impact, leading to excessive dust formation, and at the same time require a fairly prolonged drying time, the use of a direct fired rotary dryer would entail costly dust recovery equipment or might even dictate the use of a wet scrubber, in which case the recovered dust in slurry form would have to be reprocessed. In this instance assuming the feedstock could be suitably pre-formed, the choice would be a through-circulation band dryer in which the pre-formed material, evenly distributed, would pass through the dryer in a quiescent state and would probably retain its pre-formed state throughout the dryer with little or no loss of dust. If, on the other hand, the feedstock was in a moist but already fine particle form it might have to be fed to an indirect-heated rotary dryer. Crystalline products which are shown on a test plant to suffer from attrition in a pneumatic or fluid bed dryer, although they may not call for long residence times, could be dried satisfactorily in a direct-heated rotary dryer either of relatively small diameter or in an indirect-direct-heated rotary dryer in which the dimension of the annular space is small and hence the degree of impact of the material during flighting is small. The appearance of crystals is frequently of paramount importance from the commercial standpoint and extensive tests on pilot-scale dryers of the above types may have to be carried out to determine which dryer yields a product of the desired quality. Attrition or loss of surface polish may be serious when crystals are processed through pneumatic or fluid bed dryers. Similar care in the selection of a suitable dryer may apply to the drying of granules of low mechanical strength which may be found to demand the use of either a through-circulation band dryer or perhaps a roto-louvre type of dryer.

Many products which require drying before marketing are in the form of

Table 10. Selection of batch dryers based on material properties.

Material handled	*Slurries, pastes, filter cakes*	*Free-flowing powders, crystals, granules*	*Large solids, components*	*Sheets, boards*	*Textile products*
Types of dryer	Dyestuffs, pigments, sludges, precipitates, filter, or centrifugal cake	Pharmaceuticals, fine chemicals, food products, and confectionery	Pottery, electrical, automotive, and metal finishing components, Paper-pulp articles	Fibre boards, wallboards, leather, mica-paper sheets, veneers, etc.	Hanks, skeins, cheeses, cops, cakes, pirns, hosiery
Tray ovens (atmospheric)	Small batches	If non-dusting, non-toxic, solvent-free. No cross-contamination	Small productions with suitably designed supports	Small productions	Cross, vertical or through-circulation
Tray ovens (vacuum)	Only if heat sensitive, solvent-bearing, or toxic	As for slurries and pastes	Not applicable	Not applicable	Not applicable
Tray tunnels	Large batches, long drying times	Large batches	Not applicable except for pottery	Leather, asbestos and acoustic boards, medium production of others	Medium and large productions
Agitated pan (atmospheric)	Medium batches. Non-heat sensitive, non-toxic, solvent-free	Medium batches, Non-heat sensitive, non-toxic, solvent-free	Not applicable	Not applicable	Not applicable
Agitated pan (vacuum)	Heat sensitive, toxic, solvent-bearing	Heat sensitive, toxic, solvent-bearing, possibility of attrition of crystals	Not applicable	Not applicable	Not applicable

Material handled	*Slurries, pastes filter cakes*	*Free-flowing powders, crystals, granules*	*Large solids, components*	*Sheets, boards*	*Textile products*
Rotary (atmospheric)	Rarely applicable unless direct-fired	As for slurries and pastes	Not applicable	Not applicable	Not applicable
Rotary (vacuum)	Not usually applicable	Heat sensitive, toxic and solvent-bearing	Not applicable	Not applicable	Cellulose acetate
Trough (atmospheric)	← as for rotary (atmospheric) →				Not applicable
Double-cone (atmospheric and vacuum)	← as for rotary (vacuum) →				Not applicable
Infra-red	Not applicable	Not applicable commercially	Surface drying only	Surface drying only	Not generally applicable
Dielectric	← not applicable →			Plywood and veneers	Not applicable

free-flowing powders. The amount of moisture to be removed is relatively small but the final moisture content acceptable for the market may be extremely low. When the drying characteristics are good, requiring short residence times, a pneumatic dryer with material recirculation together with adequate dust recovery equipment additional to the high efficiency cyclone inherent to the unit may be very suitable. Alternatively, if dusting is not a serious problem, a continuous fluidised bed dryer could be used. Where the material is not so easy to dry to the final moisture content and a lengthy residence time is required, a continuous rotary tray dryer is likely to be the choice unless the fineness of the powder is likely to result in excessive dust entrainment in the circulated air when the material drops from tray to tray. Dust entrainment can lead to dust settlement on the steam heater coils, resulting in reduction in heat-transfer rates. Direct-heated rotary dryers are likewise suitable if dusting is not a problem but should the latter be seriously possible an indirect-heated dryer would be used.

The sensitivity of a material to temperature will have an important bearing on the choice of dryer. Where the maximum permissible material temperature is exceedingly low there may be no alternative to batch drying under vacuum conditions or, in exceptional cases, in a continuous vacuum belt dryer. For moderately low temperatures where other considerations indicate that a direct-heated rotary dryer would be suitable, this would have to be of the co-current or parallel-flow type. Pneumatic dryers are essentially parallel-flow dryers and may be suitable, although it is likely that they may have to be designed in more than one stage to secure the necessary residence time to attain the required final moisture content. Continuous rotary tray dryers and through-circulation band dryers are designed to give both adequate residence time and zonal temperature control of the product during its passage through the dryer and hence may be the most satisfactory choice in spite of their relatively high capital cost. Materials which are not heat-sensitive but which have to be dried to low final moisture contents would be dealt with in a counter-flow dryer in order to secure the maximum possible thermal economy and to reduce the size and hence the capital cost of the dryer. A counter-flow rotary dryer would be direct-heated unless the material is excessively dusty or may not be brought into direct contact with the combustion gases, in which case an indirect-heated rotary dryer would be used with the heating medium flow counter-current to the movement of the product. Where steam is the source of heat and the material, pre-formed or otherwise, is suitable for drying on a through-circulation band dryer, this would be used with steam at maximum possible pressure and preferably with means for usefully recovering heat in the condensate so that the latter could be returned to the boiler plant at the most economical temperature and pressure.

The ultimate user often wishes to install a continuous dryer to deal with more than one product. This requires considerable caution because of the danger of cross-contamination and should, if possible, be avoided. Furthermore, the drying characteristics of the several products may differ and this may require re-adjustment of the dryer operating conditions resulting in improperly dried product until the dryer reaches steady operating conditions. In general it is better to have individual dryers for each product. Should factory space and capital investment considerations dictate that a single dryer be installed, the choice must be made from the types of dryer indicated as suitable for the drying operation, and a dryer which has the maximum accessibility for cleaning, minimum hold-up areas and greatest facility for attaining steady operating conditions should be chosen. Where a pneumatic dryer is a possible choice, this is in all probability the best selection for processing more than one product as it is virtually self-emptying, is effectively purged by the large volumes of air available in the system, and rapidly attains predetermined operating conditions. Rotary dryers, particularly direct-heated rotaries, usually have a fairly considerable hold-up, take some time to empty and do not settle down to steady operating conditions very quickly, which means that an appreciable amount of material may have to be discarded or reprocessed. They must, moreover, be of simple internal design and of adequate diameter to permit internal cleaning between the introduction of different products. If the feedstock tends to stick on the internal surfaces of the dryer at the feed end, the likelihood of using this type of dryer for more than one product is remote. Continuous band dryers can be used for more than one product only if dusting is not a problem and preferably where provision is included in the design for cleaning the perforated band. If these considerations can be satisfied, a band dryer can be very suitable, as a second product can be fed to the dryer after partial or complete discharge of the previously fed and dried material. Continuous rotary tray dryers and fluid bed dryers do not lend themselves to multi product drying due to hold-up within the dryer which can be considerable and may involve some time in emptying and re-establishing correct drying conditions for a second product.

The drying of components, large solids, special shapes and prefabricated products usually demands purpose-built dryers and the foregoing considerations in the selection of a suitable dryer do not apply. Similarly, the drying of textile materials, paper, decorative laminates, wallboards, fibreboards, leather, ceramics, etc., requires specially designed dryers and the choice of type and particular design is more straightforward.

Table 11 indicates a convenient manner in which to make an assessment of the most suitable type of dryer to select based on the physical nature of the feedstock, its behaviour during the drying process, its heat-sensitivity

Table 11. Selection of continuous dryers

Liquids, pumpable slurries and suspensions	*Pastes, thick slurries, filter cakes*	*Free-flowing powders, granules, crystals*	*Fibrous products*	*Large solids and shapes*	*Sheets*
Spray Dryers, for large productions. Suitable for heat sensitive materials. Desirable spherical form of final product. Allows high initial air temperatures with low solids temperature. Drum Dryers, for small and large productions. Generally for non-heat-sensitive materials. Vacuum and reduced internal steam temperature for heat-sensitive materials. Product flaky or dusty sometimes in sheet form.	Vertical turbo-tray if feed not too wet. Long residence times. Dusty materials lead to high maintenance costs. Back-mixing and pre-forming may be employed. Direct rotary: co-current if very wet, usually pre-formed. Suitable for medium heat-sensitive materials. Counter-current for low final moisture content from medium initial moistures.	Rotary, Band, Turbo-tray, Roto-louvre, Fluid bed and Pneumatic all suitable. Long residence time: Band, Turbo-tray or Rotary. Short residence time: Pneumatic or Fluid bed. Heat-sensitive: Pneumatic or band. Subject to attrition: Band, Roto-louvre and possibly Turbo-tray.	Through-circulation Band, Pneumatic in special cases, e.g. wood-meal, wood chips, peat. Direct rotary. Tray tunnel for asbestos products, textiles, cops, cheeses, cakes. Conveyorised systems for hanks, skeins, and cheese.	Tunnels. Band. Conveyorised systems Rotating multi-tray. Infra-red for surface drying.	Discontinuous sheets: Tunnels. Band (vertical impingement). Rotating multi-tray. Dielectric. Continuous sheets: Cylinder or "can" Special conveyorised systems, usually vertical impingement forced convection. Infra-red for surface drying.

Liquids, pumpable slurries and suspensions	*Pastes, thick slurries, filter cakes*	*Free-flowing powders, granules, crystals*	*Fibrous products*	*Large solids and shapes*	*Sheets*
Direct, Co-current Rotary, for large productions from thicker slurries. May require back-mixing. Not suitable for really heat-sensitive products or excessively dusty. Vacuum band, for medium productions of heat-sensitive materials.	Indirect Rotary for very dusty material if not heat-sensitive. Band dryer with or without back mixing, usually preformed. Suitable for dusty material when pre-formed wet. Easily extended for increased production later. Fluid Bed and Pneumatic – suitable with back mixing and pre-forming.				

or otherwise, together with other special considerations which have to be taken into account. Such an assessment will usually lead to the selection of two or three most suitable types of dryer as the most satisfactory possibilities. Further evaluation of this short list should now be made by a drying test on small-scale or pilot-scale dryers of these types.

This step in the selection of a dryer is most important as it will establish not only the practical suitability of a dryer but also the operating conditions, size of dryer required and its probable thermal efficiency, as also any special problems associated with feeding the dryer and precautions which may have to be taken to reduce material losses and to produce the product in its most suitable form. Reputable dryer manufacturers generally possess such pilot units on which tests may be carried out at a nominal charge, which will be waived in the event of their receiving the order for the commercial unit. In view of the final cost of an installed full-scale dryer such an expenditure on tests is fully justified. Moreover, as the tests will establish the optimum operating conditions, it is far more economical to have these carried out than to embark on extensive experimental work on the full-scale plant with consequent losses in spoilt product, energy and manpower.

Final Selection

Such pilot-scale tests may of course establish that more than one type of dryer is technically suitable for the drying of the material under consideration. The final selection will be made on a careful analysis of questions of plant costs, installation costs, available height and works area, operating costs, flexibility to meet product quality specification changes and demand, effluent problems and available sources of energy. Formal quotations, specifications, with general arrangement layout drawings, power and heat requirements and installation charges should be obtained from the dryer manufacturers expert in the selected designs of dryer, from which the overall capital and operating costs can be compared. The result, taken in conjunction with site and commercial considerations, will lead to a final selection and the placing of the order. The supplier's terms of contract should clearly define the limits of his responsibility and that of the purchaser; omission of this can frequently lead to irritating and often costly expenditure which may not have been budgeted for.

Continuous Drying of Liquids and Gases

Liquids which remain liquid when free of moisture are dried by the use of desiccants either by chemical action, absorption or adsorption in a similar manner to methods used for gases as described in Chapter XIV. Where the liquid is composed of solids in solution or suspension it will be dried either on a film drum dryer or in a spray dryer. These types of dryer and their application are described in Chapters XI and X respectively and the considerations which will determine which type of dryer to select are discussed in Chapter X.

From the foregoing it will be appreciated that the resolving of a drying problem and the selection of a suitable drying installation are not a simple matter. Quite apart from the necessity to ascertain the drying characteristics of the product and hence the types of dryer which may be suitable, many other factors have to be taken into consideration. Either the dryer manufacturer or the purchaser, or both, must take these into account in arriving at the final solution and this calls for an objective approach to the problem by both parties. Where the dryer manufacturer supplies only one or two types of dryer this will be difficult and the selection analysis will have to remain with the purchaser after receiving alternative recommendations from manufacturers of other designs of dryer. The manufacturer of a wide range of dryers will not be biased in favour of what may in fact be a less suitable dryer taking all factors into consideration, but in order that he can make a careful and correct assessment it is most important that the purchaser should take the supplier fully into his confidence on all matters relevant to the installation of a drying plant. Experienced specialists in dryer design and supply will wish to approach the problem not only from the purely technical angle but also from the processing standpoint and will wish to have as much information as possible on processes both ahead of and after the drying operation in order to formulate a scheme most beneficial to the purchaser. The physical nature and positioning of these processes may be all-important in respect of feeding to and delivering from the dryer and influence the decision on the type of dryer likely to prove most appropriate. The quality and availability of labour, atmospheric conditions, works layout, buildings, essential service lines, and topography will have an important bearing. Finally, samples for testing should be made available in adequate amounts and should truly represent the materials to be processed; where available their physical and chemical characteristics should be accurately disclosed, if necessary in confidence. Quantitative requirements should be clearly defined and, if subject to variation in practice, the most onerous requirements should be specified if these are to determine the sizing of the commercial unit.

CHAPTER V

Tray Dryers

Forced Convection Batch-Drying Ovens

The most generally used intermittent drying ovens, whether in small cabinet or larger compartment form, usually employ trays on which the wet material is supported and which are either loaded on to trucks in tiers or are inserted directly into the dryer in which they are supported on tiers of angle-iron supports. Heated air is circulated by means of fans over the wet material and the humid air is rejected to atmosphere. The trays with their load of material remain in the dryer until drying is complete, when they are withdrawn, emptied and then recharged for drying as a subsequent batch. The design of trays and their spacing will, of course, depend on the physical form or dimensions of the wet charge.

Cross-circulation Tray Dryers

Earlier designs of tray ovens, particularly larger installations, relied on the passage of heated air, fed into the dryer at one end from a fan and steam-heated battery or combustion air heater, over several stacks of trays in series, with total rejection of the cooler humid air at the other end. The shortcomings of such an arrangement are obvious, in that there is a considerable drop in temperature of the air and increase in its humidity from the point of entry to the point of discharge. This results in great disparity in the rate of drying of material near to the points of introduction of the hot air and discharge of cooler more humid exhaust air. In consequence, it was usually necessary to withdraw and reintroduce stacks of trays progressively towards the hotter end, involving considerable labour and non-uniform drying. Modern designs of tray dryers aim at having a short distance of travel for the air, usually not more than that corresponding to two or three stacks of trays, and the recirculation of the hot drying air with partial rejection of humid air and its replacement by fresh air. Arrangements are made for reheating the air after each passage over the trays. As shown earlier, the rate of heat transfer and rate of drying during the constant rate period is constantly affected by the air mass velocity; to provide such velocities in a total rejection dryer would involve excessive heat requirements in heating up cold fresh air, whereas with internal recirculation designs it is only required to re-heat the circulated air through relatively few degrees, to heat up a much lower quantity of fresh air, and to replace heat losses from radiation and convection. By employing internal recirculation, the mass air velocity will be limited only by fan horsepower costs and any danger of disturbing or entraining the material subjected to drying. If the distance of travel of this air in a single pass is small and heater units are interposed between passes, the temperature drop of the air in its passage through the dryer can be kept extremely low, which results in greater uniformity of material temperature and rate of drying. The basic principle on which most modern tray compartment dryers are designed is therefore the use of large volumes of recirculated air with frequent reheating, together with controlled introduction of fresh air into the system and controlled exhausting of a corresponding amount of humid air. Control of fresh air and humid air can be either by manual setting of dampers or by the use of hygrometric devices during the drying cycle.

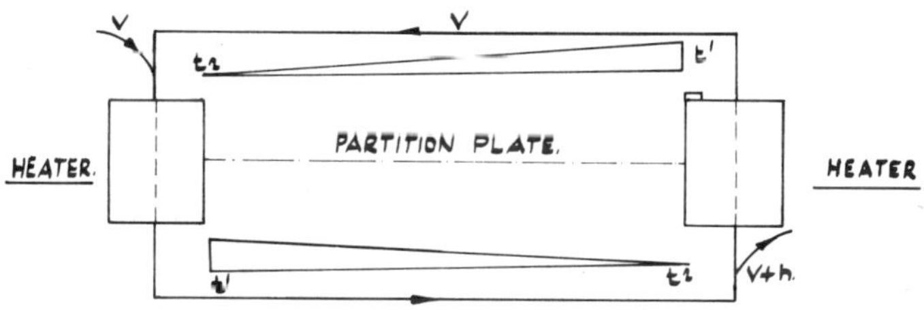

Fig. 4. Principle of recirculation dryer.

This principle is illustrated diagrammatically in Fig. 4 whilst a cross-section of a typical tray drying oven operating on this principle is shown in Fig. 8, p. 72. A large volume of air V is circulated by a fan, frequently mounted within the dryer enclosure, over a heater battery also within the dryer side compartment, and its temperature is raised to t_1. Passage of this air over the wet material will reduce this temperature to t_2, which by reheating in a second heater battery is raised again to the original temperature t_1 before re-passage over the material undergoing drying. Fresh air in smaller volume v, usually of the order of 10% of V, is introduced ahead of the heater battery, while a corresponding amount of humid air $v+h$ is exhausted ahead of the second heater. By having suitably supported stacks

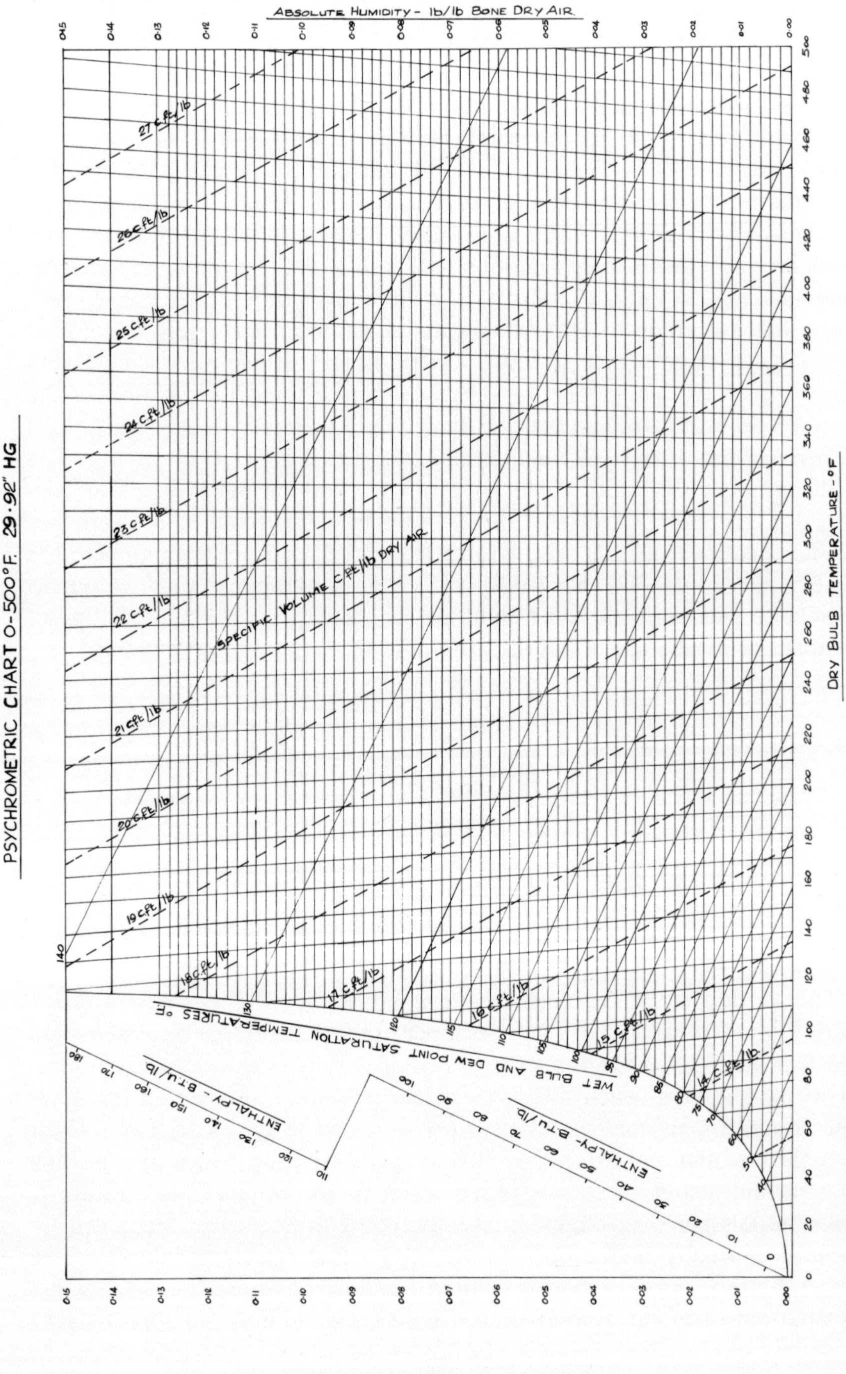

Fig. 5. Humidity—enthalpy chart.

of trays, the trays themselves act as horizontal partitions which permit this principle of recirculation. A further advantage of this method of recirculation and reheating is that drying can be carried out and completed at lower temperatures than are possible with a single passage of heated air through a tray dryer. This explains why on the introduction of this type of dryer it was found possible to replace vacuum shelf dryers to a large extent in the dyestuffs industry where, with the then existing drying ovens, it was not found possible to dry heat-sensitive products of this type other than under vacuum. That this is so can be illustrated by means of a psychrometric chart as shown in Fig. 6.

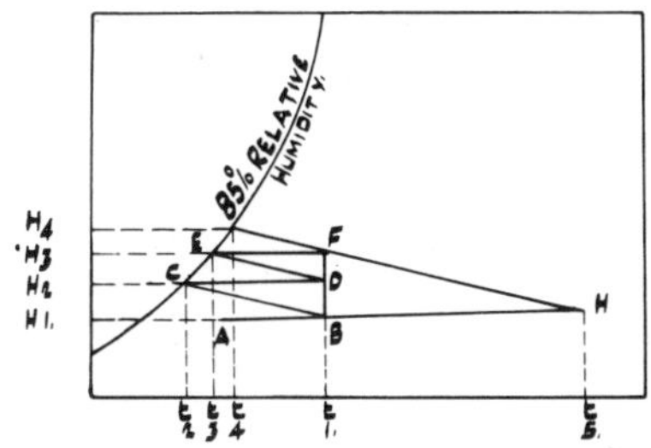

Fig. 6. Effect of reheating circulated drying air.

Fig. 6 shows how fresh air of humidity H_1 is heated in the dryer heating battery to a temperature t_1, its humidity remaining constant (line AB) and is then passed over the wet material on a tray from which it picks up moisture, increasing its humidity to say 85% relative humidity. During its passage over the material its temperature will drop, following an adiabatic cooling line, to point C on the 85% relative humidity curve corresponding to a temperature t_2 and humidity H_2. In this single passage the air will have picked up $H_2 - H_1$ pounds of moisture per pound of dry air. If this air is now reheated to the original temperature t_1 represented by the line CD and passed a second time over wet material on another tray, it will cool and pick up moisture following an adiabatic cooling line DE resulting in a new humidity H_3 and a fall in temperature to t_3. This effect can be repeated any number of times depending upon the internal design of the dryer, so that a greatly increased pick-up of moisture can be attained as compared with a single passage of air. Three passes of air, with reheating after each passage, would result in a total pick-up of moisture corresponding to $H_4 - H_1$ and a humid air temperature of t_4, whereas had this amount of drying been carried out by the same quantity of air in a single passage, attaining the same exit temperature t_4, it would have been necessary to introduce the fresh air at a much higher temperature t_5. The internal design of a compartment oven to operate on a large number of air passages and reheats is, of course, complicated and hence fairly costly as compared with a dryer designed for fewer reheats, but for large productions of thermally unstable products such as dyestuffs, and some pigments having considerable moisture contents and hence pronounced constant rate drying periods, it may justify the increased capital cost. Such a dryer is illustrated in Fig. 7. It will be noted that this principle assumes that the wet material during the constant rate period maintains air wet-bulb temperature and that drying is carried out under adiabatic conditions, neither of which assumptions is strictly true in practice.

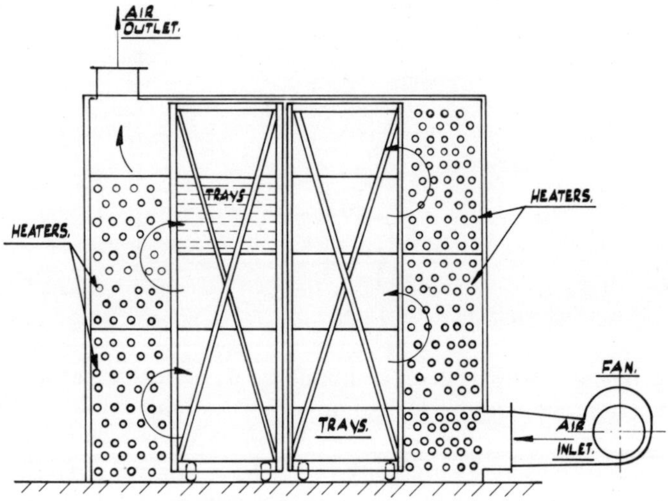

Fig. 7. Multiple—reheat circulation drying oven.

Drying ovens are usually rigidly built from rolled steel sections provided with panels and doors of double metallic sheet construction having heat insulation material between. In view of the exposed surface, thermal insulation is important, and it is customary to reduce metal-to-metal conduction by interposing heat insulation strip material between the main structure and the panel surface. Industrial sized units will usually be designed to receive trucks carrying stacks of trays whilst small pilot-scale or laboratory units will have angle slides fixed to the structure on which to support trays. Some designs are arranged with external air-circulating fans but the modern tendency is to have either propeller type or centrifugal fans built in integral with the oven structure. Heater units, whether electrical, gas-heated or steam-heated, are usually installed in the side compartments of the oven in order to conserve heat and to reduce pressure losses normally caused by external trunking. Suitable arrangements of internal adjustable louvres are employed to produce uniform distribution of circulated air whilst at the same time shielding the charge from direct radiation from the heater units which would otherwise produce non-uniform drying. Tray drying ovens are frequently subjected to very rough treatment and for this reason have to be robustly built. Although interior lining of panels and doors with sheet metal may not be desirable from the point of view of corrosion, it is usually good practice to have all external surfaces reinforced or protected by metal sheets.

The means of heating tray ovens will depend on circumstances. For pilot-scale work and small productions, heater units composed of sheathed electric heaters, thermostatically controlled, are usual. For larger commercial productions, steam heater batteries are customary where steam is available. Where gas is more convenient, direct-fired tubular heat exchangers can be used, again thermostatically controlled. Where neither steam nor gas is available and where it is permissible to bring products of combustion into contact with the material undergoing drying, direct combustion air-heating furnaces may be used. The use of the latter with internal recirculation presents certain problems and the principle of multiple reheating is difficult to apply.

The spacing of trays is usually such as to maintain low pressure losses and is generally of the order of three inches using $1\frac{1}{4}$–$1\frac{1}{2}$ in deep trays. This makes it possible to use high-volume, low-pressure propeller type fans consuming low horsepower. Air velocities up to 1,000 ft/min are quite usual between trays, although lower air velocities must be used where there is danger of entrainment of material as drying proceeds. It will be apparent that with multiple-pass tray dryers the fan capacity will decrease with the number of passes to yield the same air velocity, although as the number of passes increases there is a tendency for increased pressure losses, and hence increased horsepower requirement for a given air velocity. The most popular design employs two air passes, the middle tray constituting the horizontal partition between the two passes. With such an arrangement one single 30 in diameter propeller fan per truck or pair of trucks is adequate. Such a fan would have a capacity of about 5,000 ft^3/min and, assuming ten tray tiers above and below the centre line of the stack with $1\frac{3}{4}$ in clearance between loaded trays, would give an air velocity of 1,065 ft/min between 32 in wide trays. Were the fan to furnish air across the full cross-section of the stack of trays, the air velocity would be approximately halved.

Different designs of tray dryers are illustrated in Figs. 8 to 12. The two-pass recirculation design mentioned above is shown in Fig. 8 as applied to a two-truck drying oven, whilst Fig. 9 shows a single-pass recirculation oven with provision for reheat between trucks and additional reheat in the upper compartment. The latter design will call for greater height to accom-

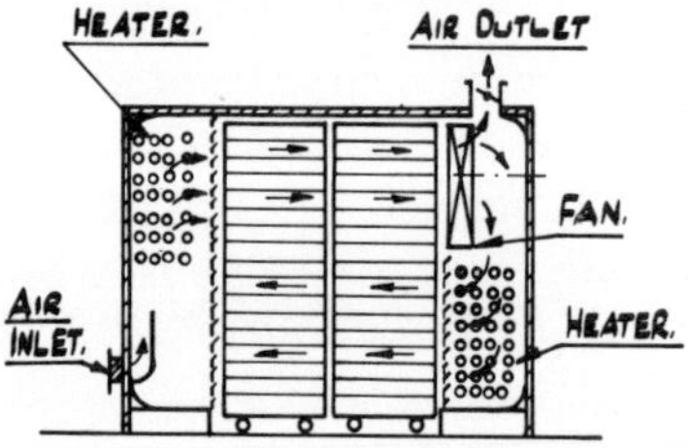

Fig. 8. Two-pass recirculation oven.

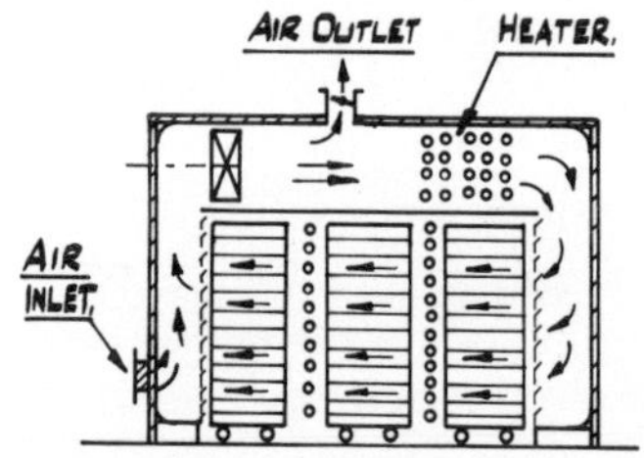

Fig. 9. Single-pass recirculation oven.

modate the oven and a higher horsepower fan to produce the same air velocity between trays than the oven shown in Fig. 8.

Uniform distribution of air within the drying oven is important and internal louvres should be adequately provided and correctly set to this end. Stagnant air pockets within the dryer should be avoided and for this reason curved deflector baffles are provided at corners. Fresh air inlets should be located ahead of the heaters and, where possible, humid air exhaust vents should be positioned in such a way as to expel air after its passage over the material held on the trays. Where extremely uniform drying of material throughout the drying cycle is important, the design of oven should be such as to permit of frequent reversal of direction of air flow.

It is not always possible or convenient to load high moisture content filter cake or pulp uniformly on trays, with the result that free air clearance between trays becomes reduced and pressure losses are increased. For such high duty performance, it is advantageous when using propeller type fans to provide two fans in series in the air circuit as shown in Fig. 10. The increased size of side compartment to accommodate the second fan can be utilised to increase the heating surface, the net result in practice being an enhancement of the evaporation rate, shortening of the drying time and lower specific heat consumption.

The air circulation ratio to give maximum thermal efficiency whilst attaining optimum drying cycles will depend upon the extent of the constant-rate and falling-rate periods. With high initial moisture content materials, a relatively high fresh air supply will be called for during the initial stages, both to promote high evaporative rates and to maintain the air below dew-point, so as to prevent recondensation on both material and the inner oven surfaces. During the falling rate period, the recirculation ratio will be maintained high at 90–95%, and it is often possible to operate with fresh air inlets completely closed, relying on small natural leakages on the suction side of the fans to sustain fresh air requirements. Regulation of fresh air and humid air inlet and exhaust dampers can be automatic

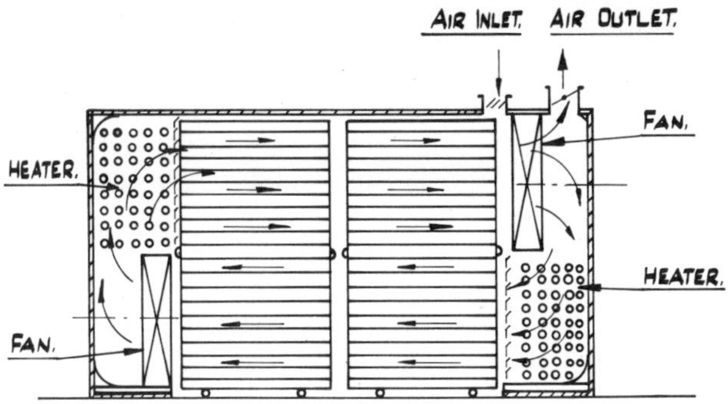

Fig. 10. High-duty two-fan, two-truck recirculation oven.

by means of suitable mechanical linkages operated from a programme controller, although it is more usual to make manual adjustments as found most suitable for the particular dryer charge. Temperature control of the circulated air is, however, usually by thermostatic instrument control, care being taken that the sensitive thermocouple is protected as far as possible from radiation from the heater unit. Where products of high purity are dried it is good practice to fit filters to the fresh air inlets and, in extreme cases, within the oven itself after each pass over the material. When noxious vapours are evolved or when corrosive constituents are present, it may be necessary to operate at unusually low recirculation rates to keep their concentration at a minimum. In extreme cases, or where inflammable or explosive volatile constituents are present, it may be necessary to adopt total rejection of the drying air.

The performance and thermal efficiency of tray drying ovens will largely depend upon the initial moisture content, the required final moisture content and the drying characteristics of the charge, together with the design of the oven and its method of operation. Materials which have to be dried to very low moisture contents and the falling rate drying rate of which predominates will usually entail lengthy drying times leading to relatively high heat losses and hence lower thermal efficiencies. High initial moisture contents and reasonably low final moisture contents will be conducive to good thermal efficiencies and high overall evaporation rates. In this connection the close fitting of panels and doors and the proper grouting-in of the oven on the works floor will have an important bearing. Rates of evaporation over the range of 0·025–0·3 lb/h/ft^2 of exposed material (tray) area are normal with specific steam consumptions for steam-heated ovens of 1·8 to 3·5 lb of steam. For well-designed multi-pass recirculation ovens correctly operated evaporation rates of 0·15–0·3 lb/h/ft^2 and specific steam consumption of 1·8–2·3 lb/lb of water evaporated are normal. Such designs of ovens, built as tunnels of considerable length and operated either continuously or semi-continuously, can result in specific steam consumptions as low as 1·5 lb/lb of water evaporated. It is, of course, assumed that the drying installation has efficient thermally insulated panels and doors and that these are a good tight fit with insulation between metal-to-metal joints.

Through-Circulation Tray Dryers

It is normal to charge the trays with wet product in the form in which it is produced in preceding processes without any special prior treatment. However, if the wet material is submitted to an extrusion or pre-forming treatment or is in granular or fibrous form, there is a chance of either drying the material normally on trays by cross-flow air circulation or adopting through-circulation on perforated or wire-mesh trays. In the former, where no departure from the above described designs of ovens is required, evaporation rates can be materially improved and drying times reduced. The loading densities will, of course, be reduced, but even so there will usually be a net gain in terms of production rates for a given size of oven. Thus an evaporation rate increase of 100% may be secured and drying completed in about one-third the time. The net gain in terms of oven out-turn of dried product may be as high as 61–70%. Adoption of a system of through-circulation on the lines shown in Fig. 11 will show an even greater improvement due to the greatly increased evaporation rates and reduced drying time.

In this design, which in other respects will be similar to a one- or two-pass recirculation oven, special sloping baffle plates are provided which

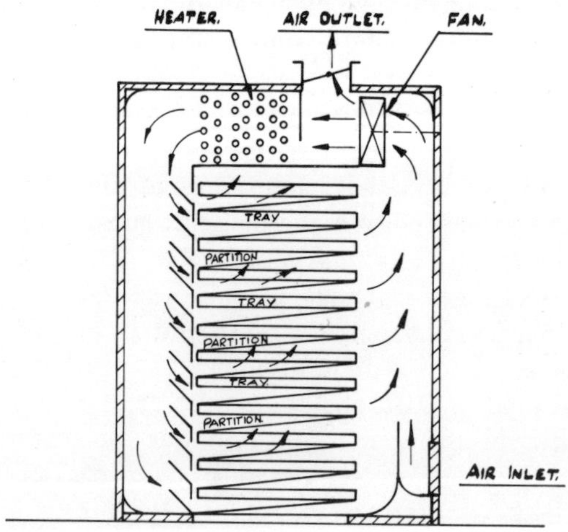

Fig. 11. Batch through-circulation drying oven.

Table 12. Forced convection tray dryer performance data

	*Coated tablets**	*P.T.F.E.**	*Aspirin base granules**	*Stearates†*	*Chalk†*	*Filter cake†*	*Filter cake†*	*Filter cake†*
Capacity, wet charge, lb	120	80	56	20,000	1,800	3,000	2,800	4,300
Number of trays	40	20	20	320	72	80	80	80
Tray area, ft^2	140	70	70	4,800	1,130	280	280	280
Depth of loading, in.	0·5	1·0	0·5	2·0	2·0	1·0	1·0	1·5
Initial moisture % w/w basis	25	25–30	15	71	46	70	70	80
Final moisture % w/w basis	Nil	0·4	0·5	0·5	2·0	1·0	1·0	0·25
Maximum air temperature, °F	113	284	122	200	180	300	200	200
Loading, lb/ft^2	0·9	1·2	0·4	0·9	0·91	3·25	3·04	11·7
Drying time, hours	12	5·5	14	24	4·5	22	45	12
Overall drying rate lb/h	2·6	5·3	0·84	62·5	185	96·6	43·2	90
Evaporative rate, lb/h/ft^2	0·0186	0·05	0·008	0·013	0·327	0·341	0·184	0·317
Total installed hp	[illegible]	1	1	45	6	4	2	2

**Courtesy Calmic Engineering Co. Ltd.*
†Courtesy A.P.V.-Mitchell (Dryers) Ltd.

lead the circulated air to the base of each tray – itself perforated or of wire mesh – through the bed of material on the trays and thence to the circulating fan intake. Here it is the superficial air velocity of air approaching the areas of the trays which matters and this will be designed around a velocity of 150 ft/min or such a figure as the material will tolerate without entrainment. By this method evaporation rates as high as 5 lb of water/h/ft^2 can be attained, but fan capacities will have to be adequate, and the drying times will have to be short enough to more than compensate for loss of loading capacity of the oven as a whole. As very important reductions of drying times are possible by using through-circulation, the feasibility of this method of dealing with granular and extruded materials is worth investigating. Pilot-scale drying tests on materials subjected to cross-flow and to through-circulation conditions will yield data as to drying time and tray-loading capacity from which an assessment can be made of the respective performance and economics of the two types in an industrial installation. Naturally, for large production rates of one product, a through-circulation band dryer of the continuous type is preferable to a batch oven.

A variety of types of tray on which to dry wet materials is available and the choice will usually depend on the physical and chemical nature of the material. Where the material form is suitable, perforated or wire-mesh trays are an advantage as they permit drying from both surfaces with a halving of the depth of material and consequent reduction in drying time. Metal trays, whether solid or perforated, are to be preferred in view of their relatively high thermal conductivity. For avoidance of contamination, vitreous enamelled trays are preferable to mild steel trays; some vitreous enamels have a reasonably high resistance to corrosion. Where corrosion is a serious factor, stainless steel, aluminium or fibreglass trays must be used. In all cases, so that the trays can be completely emptied and cleaned with ease, they should have radiused sides and corners. Tray sizes will largely depend on convenience bearing in mind the weight of a wet charge and the type of receptacle into which the material is to be emptied. A widely used size in Great Britain is 32 in long by 16 in wide ranging in depth from $1\frac{1}{4}$–2 in. Such a size is convenient for handling by either sex and the 16 in dimension will discharge into most receptacles such as kegs and barrels. Where trucks are employed in which to stack trays they should be robust and such as to give easy loading of trays; they should be provided with good quality roller or ball-bearing castors, there being one fixed and one swivelling pair. For wood floors, castors should be tyred with heat-resisting rubber. For heavily laden trucks, it is good practice to sink shallow rolled steel channels with splayed approach ends in the floor within the oven so as to eliminate wear and tear of the floor.

It frequently happens that, after installation of tray drying ovens for an initial production rate, market demands increase calling for additional drying capacity. Fortunately most dryers of this type on the market today are designed on the unit principle so that they can be extended at a later date and the purchaser should ensure at the outset that the design which he favours permits this with the minimum of down time. As such ovens are usually delivered fully assembled or in suitable sections to pass through doorways, and as no special foundations other than a level floor are required, the extension of an oven will require very little time to accomplish. Oven bases should be grouted to the floor to eliminate air leakages and where very heavy wet loads are to be dried it is advisable to provide shallow gutters for any water condensed in the earlier stages of drying to drain away. As initial condensation is likely to occur, particularly where there is evolution of corrosive vapours, regular maintenance of the interior of the oven and of the trucks by application of a heat-resistant and perhaps corrosive-resistant paint should be carried out. Likewise fan bearings should receive regular attention, noting that for drying temperatures above 240°F external fan bearings are to be preferred. Exhaust trunking connected to the humid air outlets of drying ovens should be of adequate dimensions and correctly designed to produce minimum pressure losses. Where very wet material is dried, with high evaporation rates in the earlier stages, condensate drainage launders should be provided inside the exhaust uptakes adjacent to the humid air outlets. If the exhaust from a number of ovens is dealt with by means of common trunking, a suitable condensate drain should be fitted to the uptake. Condensate drainage back into the drying ovens is a frequent cause of poor performance. Two further, if elementary, causes of poor performance are inadequate steam trapping of steam heater batteries and incorrect wiring-up of circulating fans. It is usual for the manufacturer to mark the direction of rotation of the fans and this should be checked at the time of installation.

Tunnel Dryers

Drying ovens designed on the unit principle can be extended to any convenient length or they may be purchased at the outset in tunnel form. If they are of the cross-flow recirculation type, the tunnel will be composed of a number of units each of which can be thermostatically controlled and, if desired, humidity controlled. As loading will take place at one end and discharging at the other, drying conditions can be controlled as desired throughout the drying cycle. Generally, fresh air inlets and humid exhaust offtakes will be spaced along the length of the tunnel to suit the rate of evaporation over the drying curve, the general movement of air from inlet to outlet being counter to the movement of trucks in order to promote high overall thermal efficiencies. Where reversal of cross air flow is desired, it is usual to blank-off one end of each truck so that partitions across the width of the oven are thereby produced. The length of tunnel will, of course, dictate whether the trucks be moved manually or mechanically through the dryer. For very long tunnels it is an advantage to have a neutral section at each end one truck long to avoid interference with drying conditions at each end of the tunnel proper. Automatic door opening and truck propulsion gear are often justified on large installations, and it is always good practice to carry spare trucks and trays over and above the dryer capacity in order to save effective drying time. A typical cross-flow recirculation drying tunnel is indicated in Plate XIII.

Tunnel dryer designs can, of course, take several forms depending on the method of air movement or circulation and of air heating employed. Different arrangements are shown in Fig. 12.

The design shown in Fig. 12(*a*) may be arranged for either counter-current or parallel air flow in relation to the movement of the charge and is an improvement on a purely adiabatic design in so far as provision is made for recirculating the main volume of air over the heater and the

Plate XIII. Continuous cross-flow recirculation tunnel dryer. *Courtesy Proctor Schwartz Inc.*

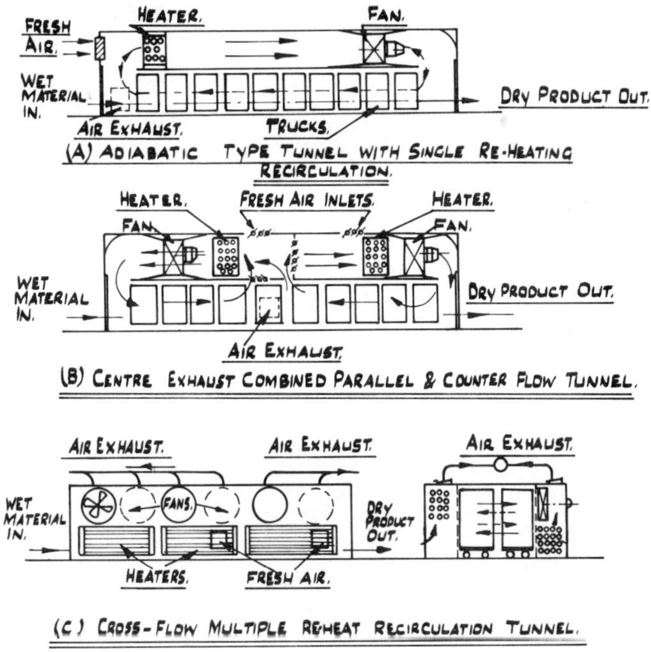

Fig. 12. Tunnel-dryer designs.

charge. However, it suffers from considerable air temperature drop from end to end, with greatly increasing humidity, thus impairing the rate of drying of the charge in the earlier stages of drying. Design (*b*) is an improvement on this in that the maximum thermal load at the charging end can be catered for independently from that part of the drying cycle which can be represented by the falling-rate period. The recycle ratio can be separately controlled and drying rates thereby improved. Design (*c*) employs cross-flow at right-angles to the movement of trucks with air recirculation and reheating at each section of the tunnel. It can therefore be operated at a constant temperature or at zonal temperatures subject to control throughout the length of the tunnel; fresh air inlets and exhaust air outlets can be provided at several points and the general movement of air from points of introduction to exit can be arranged on the basis of counter-flow, parallel flow or a combination of both. Whether the fans and heaters are located integral with the tunnel, as shown, or are mounted on top of the tunnel is a matter of preference and floor areas and height available. Greatest drying efficiencies are naturally obtained with the former arrangement.

Vertical Turbo-Tray Dryers

A design of dryer which can be regarded as essentially a continuous tray dryer in which the wet material is continuously introduced on to a series of trays and is discharged continuously in the dried state is the type usually referred to as a vertical turbo-tray dryer. This is illustrated diagrammatically in Fig. 13 and Plate XIV shows a typical industrial installation.

Apart from the truly continuous operation of this dryer, it has the advantage that the feedstock is agitated and cascaded as it moves from one tray to the next, with the result that the material constantly has freshly exposed surfaces and drying is especially uniform. This design of dryer generally takes the form of a cylindrical or hexagonal insulated vertical housing having either internal or external vertical compartments to accommodate steam heating coils, and within which several tiers of seg-

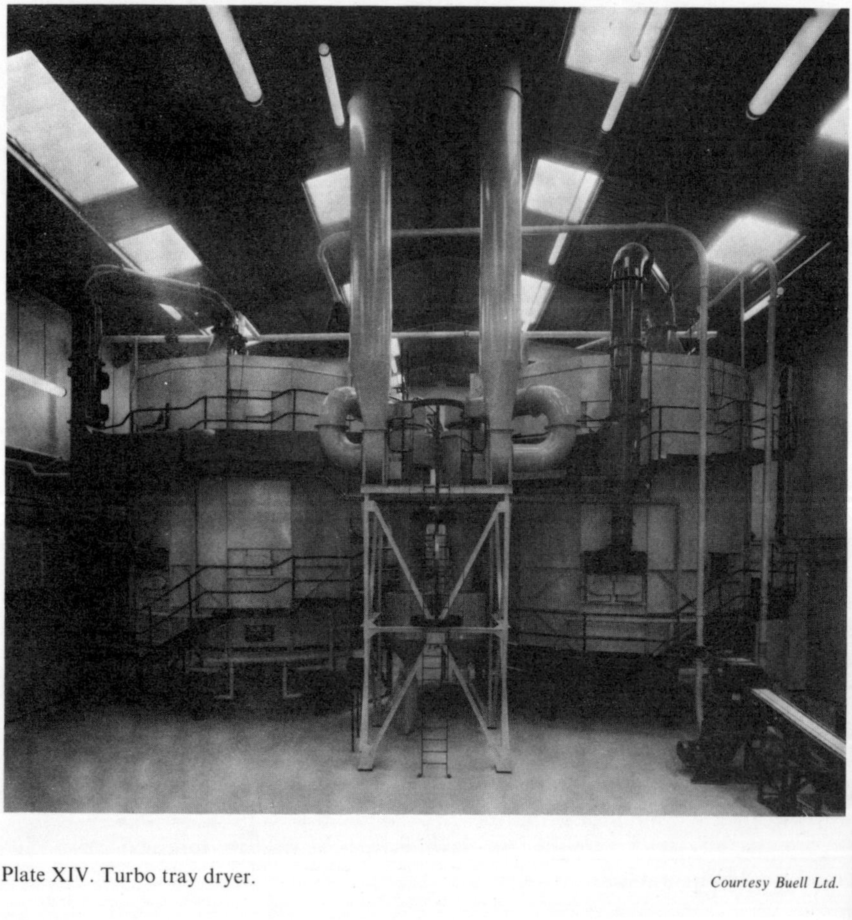

Plate XIV. Turbo tray dryer. *Courtesy Buell Ltd.*

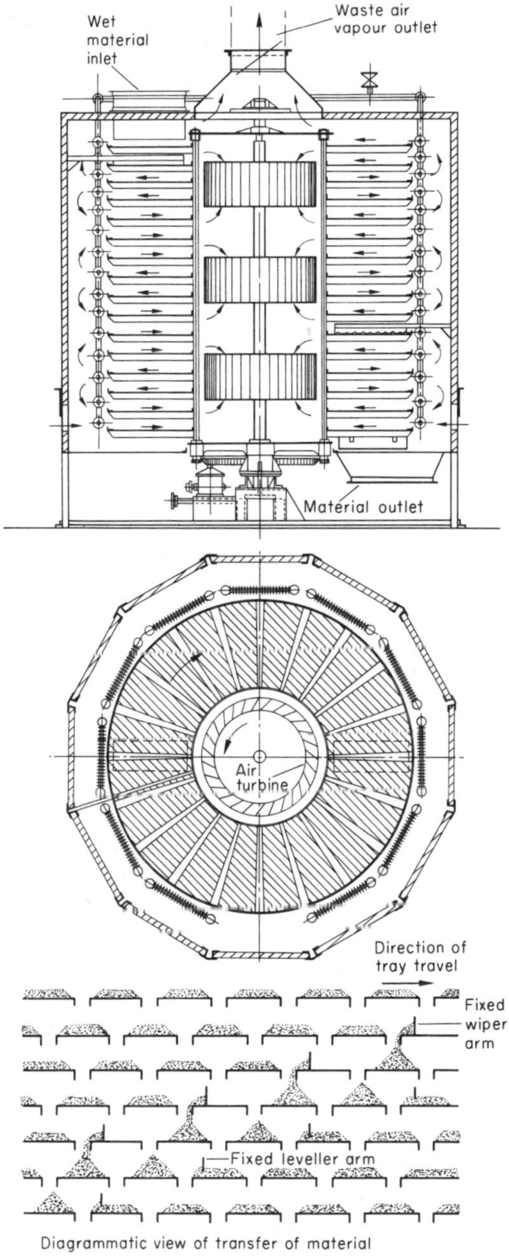

Fig. 13. Turbo-tray dryer.

Courtesy Buell Ltd.

mented trays slowly rotate. The central area houses a vertical driven shaft carrying a number of open-vaned centrifugal fans which promote air flow over and above trays and through the heating coils. In direct-fired dryers of this type, the heating coils are omitted and hot gases are introduced into the internal circulation system at one or more inlet ports up the height of the dryer. Wet feedstock – usually of a reasonably friable or free-flowing nature – is fed on to the top tray on which it is levelled by a stationary blade or plough. Each tray has a slot extending its radial width between each segment and a secondary stationary plough discharges through this slot the contents of each tray to one at the next lower level. The feedstock thus progressively moves from tier to tier to the base of the dryer where it is discharged by a screw conveyor or other means. Where desirable, the lowest tiers of trays may be supplied with cold air when they act as a cooling section. Air velocities will usually be comparable with those used in cross-flow batch tray dryers and are controlled to suit the material being dried by adjustment of the speed of revolution of the fans. It is usual in steam-heated turbo-tray dryers to preheat the incoming fresh air before it enters the base of the dryer; humid air is generally exhausted at the top of the dryer, which is operated as a continuous counter-flow dryer with multiple reheating of the circulated air.

During its passage through the dryer, the feedstock generally remains quiescent on the trays, being disturbed only for discharging from tier to tier. Dust formation is not normally serious except in the case of very fine light materials where there is the possibility of dust accumulation on steam heater coils, which in consequence have to be frequently cleaned by compressed air blowing in order to maintain good heat transfer rates. It is rarely necessary to employ auxiliary dust recovery equipment on the exhaust gases. By virtue of the very large number and area of trays which can be installed in large tall dryers of this type, the productive capacity for a given factory area is considerable. Properly designed, the power requirements are low due to the low fan speeds employed and the very moderate speed of rotation of the trays—usually one to one tenth of a revolution per minute. Thermal efficiencies are high and maintenance cost – except where corrosion or excessive dust formation occurs – is low. Large installations usually call for special and somewhat costly excavation and foundations but moderately sized and small units can be mounted on a factory floor. Typical performance data for this type of dryer are given in Table 13 and a steam balance diagram is given in Fig. 13(*a*).

Heat Losses

The thermal energy required for the drying of wet solids can be considerable and in the case of very wet feedstock may represent an important propor-

Table 13. Vertical turbo-tray dryers performance data.

	*China clay**	*Bread crumbs**	*Cu-Ni concentrate**	*Catalyst pellets†*	*Kaolin†*	*Calcium chloride†*	*Urea†*	*†Vitamin powder†*
Dryer height	—	—	—	23 ft	23 ft	47 ft	47 ft	12 ft
Dryer diameter	—	—	—	19 ft	19 ft	31 ft	15 ft	9 ft
Tray area, ft^2	7,000	2,000 (drying) 1,000 (cooling)	2,900	—	—	—	—	—
Capacity, lb/product/h	31,000	1,680	19,000	4,200	10,000	24,000	5,000	200
Initial moisture, % w/w basis	30	36	22	45	35	25	20	20
Final moisture, % w/w basis	10	5	5	18	5	1	0·2	5
Product temperature, °F	160	100	200	—	—	—	—	—
Residence time, min	40	40 (drying) 20 (cooling)	25	—	—	—	—	—
Evaporation rate, $lb/ft^2/h$	9,100	804	4,060	2,050	4,600	11,000	100	37
Method of heating	External oil	Steam	External oil	External gas	External oil	Internal gas	External steam	External steam
Heat consumption, Btu/lb, moisture evaporated	1,750	—	2,200	1,750	1,850	1,800	3,500	2,700
Installed hp	80	25	60	23	47	65	75	$2\frac{1}{2}$

*Courtesy Buell Ltd.
†Courtesy The Wyssmont Company Inc.

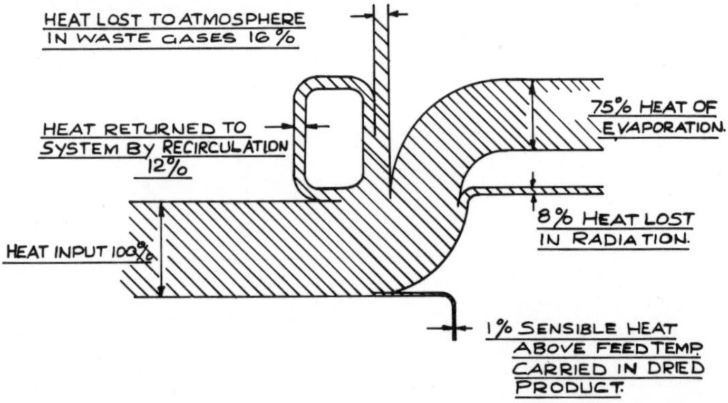

Fig. 13(a). Heat-balance of turbo-tray dryer. *Courtesy Buell Ltd.*

tion of the cost of manufacture. High thermal efficiencies as a result of correct dryer design and operation must therefore be aimed at. Drying ovens present considerable exposed surface areas with the result that heat losses may be high if attention is not paid to adequate thermal insulation. With good panel and door insulation, elimination of metal-to-metal contact and the use of light coloured paint or low radiation external metal panels, considerable heat losses can be avoided. Such heat losses may vary between 0·1–0·6 Btu/h/ft^2/deg F temperature difference between the inner and outer panel surfaces depending upon the quality, thickness and design of insulated panels. It is good practice to have 1 in thick good quality thermal insulation for drying temperatures up to 200°F increasing by half an inch for every increase of 50° in drying temperature for batch drying ovens, increasing the thickness by 50% for large tunnels and when the drying time of the wet material is very considerable.

The heat consumed in raising the temperature of trucks and trays and of the material itself to the final drying temperature is, of course, not amenable to control, but that required to heat up the fresh air to the temperature at which it is finally exhausted from the system can be the source of considerable heat loss as sensible heat in the exhaust air. This does not mean, however, that exhaust air temperatures should be kept as low as possible without reference to the humidity of the air. The heat lost in the exhaust air although directly proportional to the rise in temperature between that of the ambient fresh air and that of the exhaust air is inversely proportional to its increase in humidity. As the absolute humidity rises more rapidly than the temperature rises for a stated percentage saturation, higher exhaust temperature at high absolute humidities may lead to more efficient operation. For this reason it is desirable to adjust the humid air exhaust dampers on batch ovens in stages during the drying cycle either manually or by a programme controller, and to set exhaust dampers correctly at several points on a drying tunnel. It might be argued that a worthwhile economy would result from the use of a heat exchanger in the exhaust air with which to preheat the fresh cold air introduced into the dryer. However, the heat-transfer coefficient of such an air-to-air exchanger is so low that only when exhaust temperatures are extremely high – rarely met with in this type of dryer – would the size and capital cost of such an exchanger be other than excessive. For large tunnels which are steam heated, heater battery condensate-to-air heat exchangers on the fresh air supply are a better proposition and could lead to a saving of 10–20% of the total heat consumption.

Batch Vacuum Tray Dryers

Batch vacuum shelf or tray dryers are today generally used for relatively small productions of materials which are excessively heat-sensitive, subject to oxidation by air or which are toxic or contain toxic or inflammable volatile components. They are essentially conduction dryers which rely on the evaporation of moisture under reduced pressure by conduction of heat, usually at very moderate temperatures from steam-heated platens or coils which support trays of the wet material or by circulated fluid heating medium; in small units, heating is sometimes by electricity. They consist of a cast iron rectangular chamber or a steel cylindrical chamber, fitted with a vacuum-tight charging door, into which the heated shelves are fitted, and provided with a steam supply and condensate drain or an electrical supply. An adequately dimensioned vapour outlet leads to a condenser, usually water-cooled, but sometimes by a supply of other refrigerant, and a vacuum pump. The charging and discharging door is either of the quick-acting type or is provided with several wing-nuts and swivel bolts. With this type of dryer, a complete complement of loaded trays is introduced into the dryer and is removed at the end of the drying cycle. A more modern design which permits semi-continuous operation is provided with a series of self-sealing independent trays which can be introduced and withdrawn from individual compartments inside the vacuum chamber whilst the remaining trays are undergoing drying. They can therefore be charged and discharged consecutively without interruption to the drying of the other trays and, in consequence, have a considerably enhanced production capacity as compared with older more conventional types. Such a dryer is shown in Plate XV.

This vacuum tray dryer illustrated has such original features in its design and operation that it merits more detailed description. Invented by A. L.

Plate XV. Drawer-type vacuum shelf dryer. *Courtesy Calmic Engineering Co. Ltd.*

Andersen of Copenhagen, it is manufactured in Great Britain and certain other countries by Calmic Engineering Co. Ltd. and its foreign associated companies. Each of the trays in this dryer is jacketed and may be heated by a circulated liquid medium, steam or sub-atmospheric steam. Heat-transfer rates, particularly when normal or sub-atmospheric steam is used, are consequently high; with steam, heat transfer can be up to ten times higher than with hot liquid heating. When vacuum or sub-atmospheric steam is used high heat transfer rates are obtainable at the reduced temperature of the steam, and these temperatures can be closely controlled by controlling the degree of vacuum at which the steam is maintained. Table 14 gives properties of dry saturated steam at depressed pressure and it will be seen that the total heat in the steam remains substantially constant over the pressure range whereas the latent heat increases with increase in degree of vacuum.

When sub-atmospheric steam is used as the heating medium, it is usually injected into a cylindrical compartment beneath the dryer in contact with a quantity of water and the desired degree of vacuum is imposed from the main dryer vacuum system. The water temperature is thermostatically controlled to the maximum which the material undergoing drying is permitted to attain and the vacuum steam passes through a pipe system to

Table 14. Properties of sub-atmospheric steam

Pressure in vacuum	*Temperature of boiling °F*	*Heat in water Btu/lb*	*Latent heat of steam Btu/lb*	*Total heat Btu/lb*
29·0	77	45	1,050	1,095
27·0	114	82	1,029	1,111
25·0	133	101	1,018	1,119
20·0	161	129	1,002	1,131
15·0	179	147	991	1,138
10·0	192	160	983	1,143
5·0	203	191	976	1,147
Atmos.	212	180	970	1,150

the tray jackets through a special valve. Steam from the tray jacket, together with condensate, is ejected through a vacuum offtake valve connected to the main vacuum compartment of the dryer which is at a lower pressure than the tray jacket. The ejector in the vacuum valve mouthpiece is fitted with a replaceable orifice, the size of which controls the amount of steam used in each tray. Each tray compartment is provided with a quick-acting spring-loaded sealing arrangement which permits removal and replacement of each tray without disturbance of the other trays. All the trays can therefore be used in rotation and the dryer is operated in a semi-continuous manner. A simple self-actuating non-return valve at the injector to each tray prevents breaking of the vacuum in the vacuum source lines on withdrawing a tray.

When operating with a heating liquid the latter is circulated at a controlled temperature to each tray jacket, heat being supplied by means of a steam or electrically heated heat exchanger below the dryer. In the case of both steam and liquid heated dryers the degree of vacuum maintained above the charge is accurately controllable and the released moisture or solvent may be condensed in a suitable condenser. Typical performance figures for this design of dryer are given in Table 15.

The external surfaces of vacuum shelf dryers should be adequately insulated to prevent internal condensation; some designs have steam-heated or electric tracer-heated chambers for this reason. It is, of course, essential to maintain door gaskets in good condition and as these vacuum joints have to be frequently broken in order to withdraw the charge this, together with the necessity for maintaining the vacuum pump in a state of high efficiency, can result in high maintenance charges. As has been shown in this chapter, modern internal recirculation re-heat tray dryers are capable of drying many materials at moderate temperatures and, unless unavoidable, it is not customary to use vacuum shelf dryers, which are costly in relation to output.

Table 15. Vacuum drawer type tray dryer performance data

	Soluble aspirin	*Paint pigment*	*Ferrous glutinate*	*Ferrous succinate*	*Lithium hydroxide*	*Tungsten alloy*	*Stabilised diazamin*
Capacity, wet product, lb/h	44	30·5	41·6	52·5	36·8	12·8	4·6
Tray area, ft^2	108	108	108	108	54	215	172
Depth of loading, in	1	2	0·5	1	1	0·5	0·75
Initial moisture, % w/w basis	72·4	49·3	25	57·4	59	1·6	22·2
Final moisture, % w/w basis	1·25	0·75	0·5	18·8	0·9	Nil	0·5
Max. temp. °F	104	158	203	203	122	239	95
Loading, lb charge (wet) ft^2	6·1	102	2·3	1·94	3·08	7·16	1·22
Drying time, hours	15	36	6	4	4·5	12	48
Overall drying rate, lb moisture evaporated/ft^2/h	0·293	0·14	0·11	0·11	0·034	0·013	0·0058
Total installed hp	6	6	6	6	3	2	5
Vacuum, in Hg	29·5	28	27	27	27	29	22–23

Courtesy Calmic Engineering Co. Ltd

As vacuum shelf dryers are composed of a series of loaded trays supported by a heating surface, heat transfer will not only be through heat conduction from the heating surface to the tray but also by radiation from the heated surfaces to the tray contents. Depending upon the temperature of the heated surface, the radiation factor may be equally as important as conducted heat. The material will undergo a normal drying cycle which will probably consist of a constant rate period and a falling rate period. As with most indirect dryers, the material will attain approximately the boiling point temperature of the moisture during the constant-rate period, gradually rising thereafter to the temperature of the heated surface. Drying rate will therefore be quite high during the constant-rate period, although this will be affected by the depth of loading on the trays, increasing rapidly with reduction in charge depth. The higher the initial moisture content, the higher the evaporation rate for a given temperature of heating medium, although average evaporation rates will be of about the same order of magnitude as for convection tray dryers.

Thermal efficiencies may be quite high ranging from 60–80% steam requirements for drying (excluding any required for thermal insulation) being 1·3–2·2 lb/lb of moisture evaporated.

CHAPTER VI

Through-circulation Continuous Dryers

Band Dryers

Just as material, in a suitable physical form, may be dried batchwise in through-circulation drying ovens, where the production rate justifies installing a continuously operating dryer this can take the form of a continuously propelled perforated or wire-mesh band on to which the wet feedstock is fed, and which passes continuously through a drying chamber. Such a dryer operates on the principle of blowing heated air through a moving permeable bed of material, the direction of movement of the air being vertically through the bed in either a downward or upward direction, or both. By virtue of the considerable exposed surface of the material and the high mass velocities of heated air through the bed, drying rates will be high. Additionally, because of the relatively shallow depth of material on the supporting band or apron, the distance through which the air passes is short, varying from one inch to ten inches usually, and drying is, in consequence, extremely uniform. Furthermore, such a moving bed is in a quiescent condition so that the danger of attrition to fine particles or dust is substantially absent. Not all feedstocks are initially in a physical form such as to present a permeable bed but the use of pretreatment devices such as extruders or expellers, which are discussed later, will usually attain the desired physical state. Fibrous, flaky, granular and crystalline materials will generally not require such mechanical pretreatment and can be fed continuously and uniformly on to the conveyor band by means of suitable feeding devices. A dryer of this type in its simplest form is shown diagrammatically in Fig. 14 and in Plate XVI.

The essential features of a continuous through-circulation band dryer are the feeding arrangement, band or apron construction, sealing devices, air circulation and heating systems, the humid air exhausting system, the thermally insulated enclosure, access facilities and dry product discharge

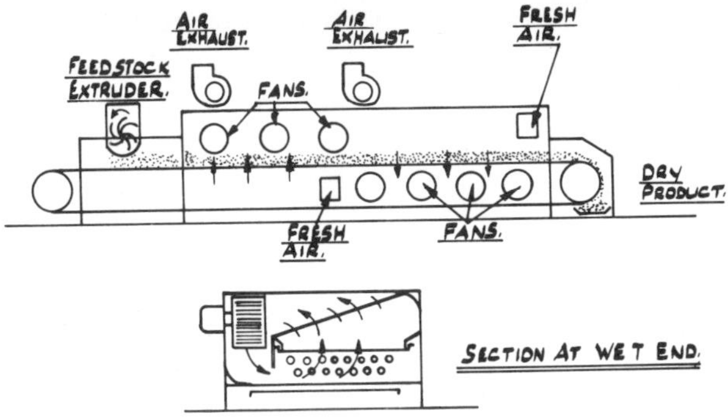

Fig. 14. Sectional view of a continuous single-pass band dryer.

Plate XVI. Continuous single-pass band dryer. *Courtesy A.P.V.-Mitchell (Dryers) Ltd.*

arrangements. There may be additional special refinements such as band washing or cleaning devices, provision for turning over the bed at one or more stages during drying, anti-fire precautions and multi-point thermostatic control. Suitable mechanical driving gear will have to be provided for the band and this will usually take the form of a variable speed-controlled drive. Band dryers can be single-pass, i.e. a continuous perforated apron running the entire length of the dryer; multiple-stage single-pass, each stage of the apron discharging to the next and so disturbing the charge with a view to presenting new surfaces to the drying medium and securing more uniform drying; or multi-pass in which a succession of aprons are installed one above the other thus reducing the length of the dryer at the expense of height. There is also the special case of the multi-pass falling-tray design in which each apron is composed of adjoining hinged sections arranged and actuated in such a way that both runs of each conveyor band are charged with material. Most modern band dryers are designed to operate on the principle of sectional or zonal internal recirculation of heated air similar to that employed on modern tray dryers, with reheating in the side compartment – usually that in which the fans are mounted – or by means of reheating coils located above or below the band. The general movement of air from fresh air inlets to humid air outlets may be either co-current or counter to the movement of the feedstock. With individual heaters in each section or zone temperature conditions can be controlled to suit the drying characteristics of the feedstock and the heater batteries in steam-heated dryers or fuel input in combustion gas heating are subject to thermostatic control. Air movement through the band and its charge may be upwards, downwards or in both directions at different points in the dryer. Typical arrangements are shown in Fig. 15, (a) being downward air flow with the steam heater battery above the band, (b) being upward air flow with the heater battery in the fan side compartment and (c) being a section of a dryer fed with combustion gases from an external source, air movement being downwards through the band.

Where the direction of air flow is changed in this manner, it is usual to employ upward air flow at the wet end of the dryer and downward air flow at the dry end so as to eliminate or reduce any tendency towards entrainment of fines or dust. The charge bed itself acts rather like a filter; any remaining entrained dust is usually filtered out when passing through the material at the wet end. When dust entrainment is anticipated, it is good practice to locate the heating coils in the side compartment where air velocities are at their highest. Heater coils will often be located above the band at the wet end, where dust is least likely to occur, so as to derive the benefit of radiant heat on to the wet material. Fresh air is introduced at the suction sides of the circulating fans at one or more points down the length

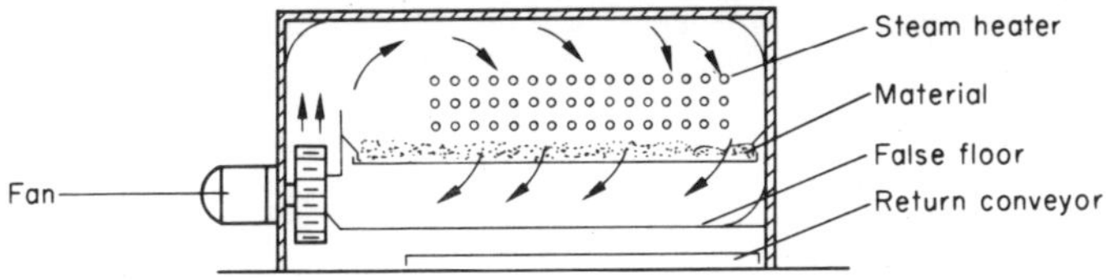

(a). Downward air flow heater above conveyor.

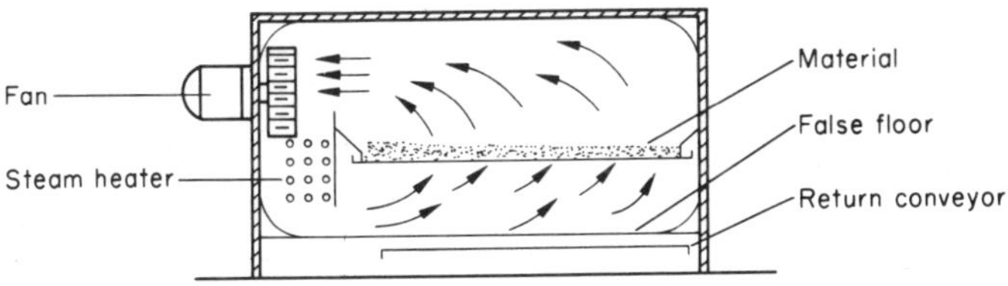

(b) Upward air flow heater in side compartment.

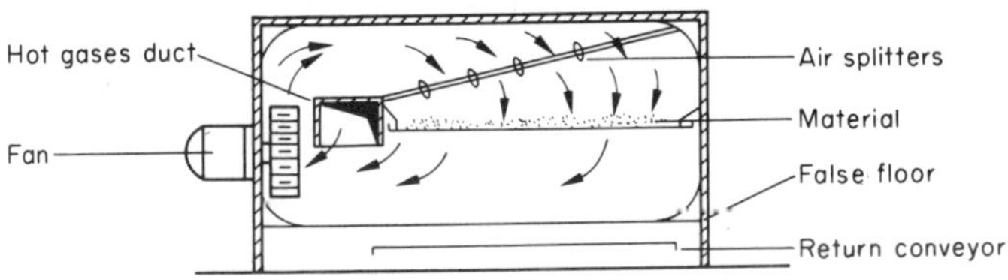

(c) Combustion heated down-draft dryer section.

Fig. 15. Section diagrams of through-circulation band dryers.

of the dryer, while humid air is exhausted, generally by extraction fans, at one or more points ahead of the heater coils.

Continuous through-circulation band dryers are particularly applicable for large productions of materials which are initially granular, flaky or fibrous or by suitable preforming can be produced as a permeable bed if drying cycles are relatively lengthy. They are especially appropriate for materials which are friable or are subject to attrition in other types of dryer. Once the material is fed on to a straight-through single-band dryer it remains on the band in a quiescent state during its progress through the dryer. High moisture filter cakes of very small particle size if passed through a suitable extruder will produce extrudates which will lie on the band undisturbed and will be discharged in their dry condition with little or no breakdown to dust and fines. By virtue of the considerable surface areas exposed and the small dimensions of the extrudate pieces, together with the high mass velocity of the hot gases passing through relatively shallow depths of material on the band, enhanced drying rates are attainable on materials with slow drying characteristics due to internal resistances. With feedstocks which undergo a considerable shrinkage during drying, this can be compensated for by employing more than one stage of perforated band running at progressively reduced speed, thereby maintaining—or even increasing—the bed depth during passage through the dryer. Such an arrangement is shown in Fig. 16.

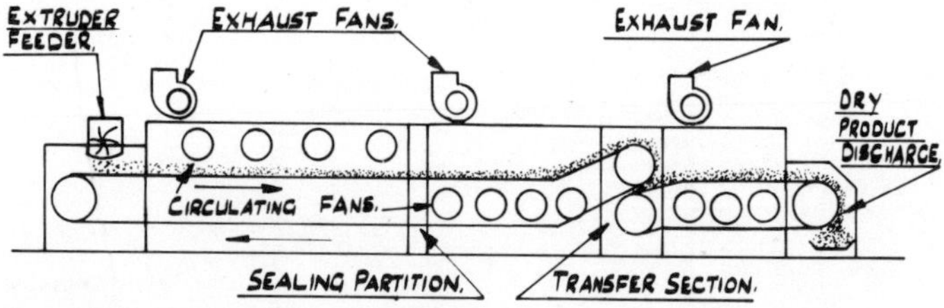

Fig. 16. Two-stage straight-through band dryer.

Feeding Arrangement

Materials which do not need any pre-forming, such as crystals, granules, flakes, fibrous or coarse materials generally which are free-flowing, will be fed on to the band by the most appropriate method to yield a uniformly distributed bed of as uniform a depth as possible. Such devices may take the form of vibrating feed tables, oscillating vibrating chutes or conveyor bands, hoppers with helical screw distributors or, where pneumatic conveying is used, solids–air separators running the full width of the band or furnished with screw distributors at their discharge. For fibrous materials, spiked drums or conveyor belts, with or without mechanical openers, are employed. When the material is initially pasty, such as filter cakes, one of many methods of pre-forming is employed, as it is essential that a uniform permeable bed be charged to the dryer. The most common method is to use an extruder of one type or another but where the material is not amenable to extrusion – usually when it is excessively wet or is thixotropic – it can be preformed on a steam-heated grooved or finned drum of adequate width. Such a feeder is shown in Plate XVII.

Grooved Drum Pre-dryer

The degree of pre-drying obtained by the use of a grooved or finned drum pre-former will depend upon the steam pressure used and the initial moisture content of the feedstock, as will the rate of evaporation per unit area of grooved drum. Thus a feedstock containing 60% moisture (w/w basis) may be pre-dried to 50% whereas an initial moisture content of 35% can be reduced to about 25–26% with evaporation rates of 26 lb/ft^2/h and 13·5 lb/ft^2/h, respectively. The use of this type of pre-former makes

Plate XVII. Finned drum preformer. *Courtesy Krauss-Maffei-Imperial G.m.b.H.*

possible the application of a continuous band dryer and considerably reduces the evaporative load and hence the size of dryer installed.

The wet material is fed on to the steam-heated drum, either directly from a hopper or through the agency of rollers or a continuous spring-loaded retaining endless blanket, and fills the spaces between the fins, the amount and dimensions of the material being controlled by the dimensions of the grooves between the fins.

Due to the heat supplied by the steam fed to the drum, a measure of pre-drying takes place, and a suitably designed doctor knife or comb removes small sticks of pre-dried material from the grooves, discharging them on to the moving band as a uniform bed, the depth of which is controlled by the relative speeds of the drum and band. Where neither this method nor the use of an extruder is satisfactory, and where the material is discharged from a continuous vacuum filter—particularly string or chain discharge filters—the use of scoring knives may produce suitably sized pieces which yield a permeable bed.

Extruders

The most usual method of pre-forming material is by the use of extruders, which may take several forms. One of the simplest types is that which consists of an oscillating drum free to rotate on its axis and which bears on a perforated curved plate through which the material is forced by pressure imposed by the drum. The material from this type of extruder, in common with most other designs, takes the form of spaghetti-like extrusions which form coils and serpentine shapes on the moving band. When the material is relatively low in moisture content the extrusion often breaks up into short pieces. Fig. 17 shows a variety of extruder designs in diagrammatic form whilst Plate XVIII shows an extruder in action.

The diameter and shape of the extrusions and the number of rows extruded will depend on the perforation size and pitching of the perforations in the extrusion plate, which will be arranged to suit the nature of the

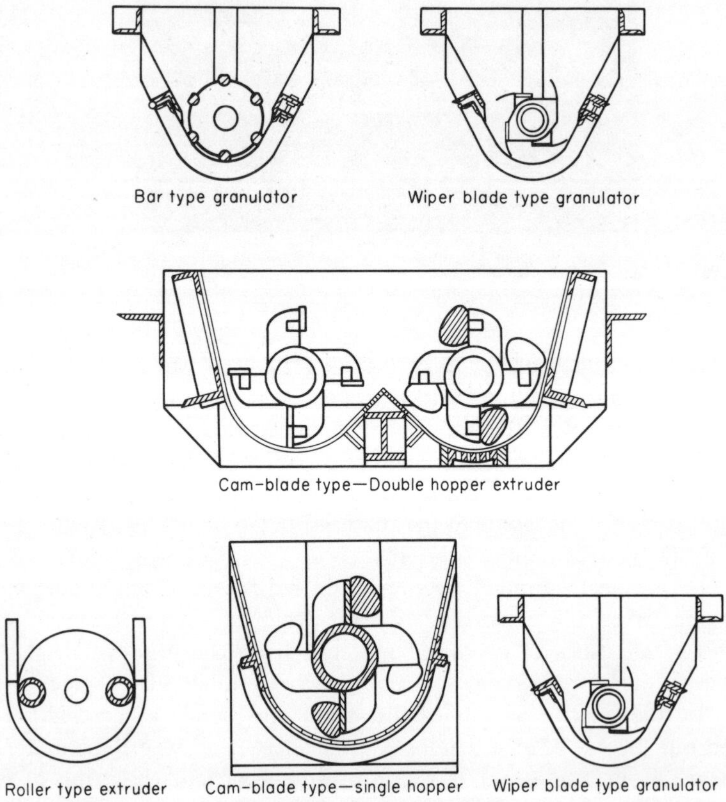

Fig. 17. Extruder and granulator equipment for band dryer.
Courtesy A.P.V.-Mitchell (Dryers) Ltd. and C. G. Sargent Sons Inc.

Plate XVIII. Continuous extrusion on to a band dryer. *Courtesy Krauss-Maffei-Imperial G.m.b.H.*

material, the rate of production required and the rate of extrusion possible. Roller extruders often have the rollers rubber-covered whilst wiper blade and cam blade extruders will have bearing surfaces in suitable metal depending on the material extruded. The rolls on the rotary roll design usually themselves rotate and are carried from a central rotating shaft. Roller and wiper blade extruders have their pressure exerting surfaces extending the full width of the trough which covers the width of the band. The cams in a cam blade extruder are carried from robust arms, there being several sets of these staggered along the central driven shaft, the trough and extruder plate again covering the width of the band. Wiper blade extruders are eminently suitable for thixotropic materials and materials which should not be subjected to too much pressure. They are also used for material which will granulate easily. Roller extruders will generally meet the more usual extruding duties whilst cam blade extruders will be used for materials more difficult to extrude and requiring considerable pressure. One design of extruder very suitable for pigment and chemical filter cakes is designed on a gear pump principle, there being twin-driven gear toothed members extending the length of the feed trough between which the filter cake passes and is forced through a nozzle plate which is hinged for ease of cleaning and maintenance. The perforated band of the dryer must have an adequate external length to receive the material fed to it by whatever means and this may be open or encased by an extension to the main dryer body, in which case it is usual to have an inspection window above the band in order to observe the performance of the extruder. Where free moisture is liberated through the action of the extruder, it is usual to provide a drain pan below the upper run of the band with a run-off drain.

Band Dryer Design

The dryer body will be built up of easily removable insulated panels and hinged insulated access doors. Complete accessibility is important for purposes of cleaning and maintenance. As with batch-drying ovens it is good practice to provide heat insulation between doors and panels and the main dryer structure to reduce heat conduction to the outside surfaces. Thickness of panel and door insulation will be generally similar to that employed with batch ovens, but in view of the considerable exposed areas it is unusual for it to be less than $1\frac{1}{2}$ in. The panel to structure insulation should be such as to give air-tight joints, in view of the suction and pressure developed by the centrifugal circulation fans usually employed. Similarly, suitable and effective hinged sealing dampers or seal-plates are fitted at the entry and exit and, in multi-stage dryers, between each stage. The delivery end of the dryer conveyor band is usually totally encased, with a glazed hinged inspection panel. Discharge of dried product will be by means of a transverse screw or band conveyor or by chute to a lower floor.

Several designs of dryer bands – sometimes referred to as "aprons" – are used. Wire mesh aprons are today rarely used except when dealing with textile raw materials such as loose wool, cotton, etc. Wire mesh bands tend to distort and disintegrate in use and are subject to excessive stretching. Normally aprons composed of individual trays or "flights" carried from a pair of matched roller conveyor chains are employed, the actual design of flight largely depending on the manufacturer's standards and the load to be carried. For light loads they will be flanged only, but for heavy loads they will be provided with one or more substantial stiffening girts. Simple flanged flights may merely butt, one against the other – in which case accuracy of folding is essential – or in common with other types will be hinged about hinge-pins extending the full width of the apron. Again accuracy of manufacture is important to ensure free movement at the hinge but no excessive clearances which may lead to fall-through of fines. The actual design of perforation will depend upon the material being dried and will range from round $\frac{1}{16}$ in perforations to elongated narrow slots usually of a maximum size of $\frac{3}{32} \times \frac{1}{2}$ in. It is usual to leave one or two inches of unperforated metal at each end to accommodate a stationary sealing guide supported from the dryer sides. For deep beds of material it is usual to fit pressed metal moving stock guides at each end of the flights which engage with adjoining guides so preventing material being deposited on the conveyor chains. Flights, stationary and moving stock guides are made in mild steel or corrosion-resisting alloys to suit the material processed. Where moving stock guides are not used, it is not unusual to provide the stationary stock guides with a sealing strip composed of brake-lining or other suitable material bearing down on the unperforated ends of the flights. Where there is a tendency for the product to adhere or blind the flights it is customary to fit rotating cleaning brushes and even water sprays acting on the return run of the apron. Fig. 18 illustrates typical flight designs and flight sealing arrangements.

Some materials, particularly when bed depths are considerable, tend to coalesce to some extent and form a continuous semi-solid blanket of material which does not easily break up and disengage itself from the band

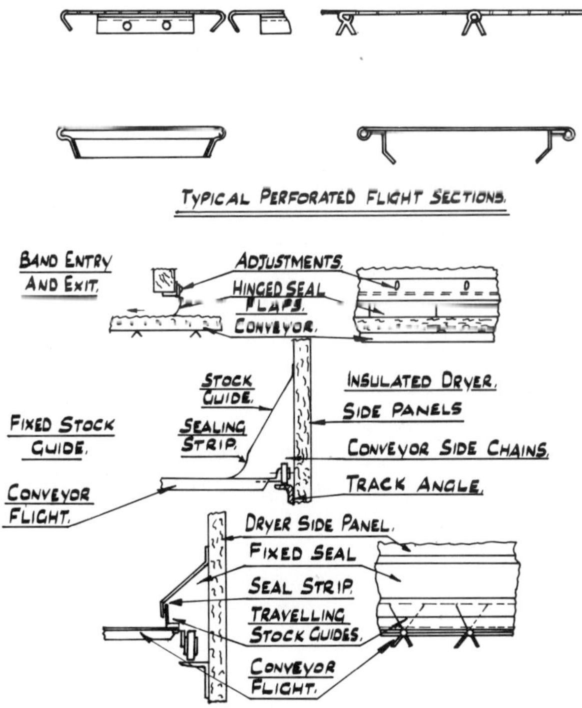

Fig. 18. Flight designs and sealing arrangements.

at the discharge end. Several designs of breakers or beaters are available to overcome this difficulty and are generally independently driven, although on smaller dryers they may be driven from the apron drive shaft through suitable gearing. They will usually be designed and counter-balanced so that they follow the movement of each flight as it changes its direction over the discharge chain sprockets. In the drying of synthetic rubber and synthetic resins such devices are generally essential. They may be necessary at each stage of a multi-stage straight-through dryer and in multi-deck dryers. A single-stage straight-through band dryer may also be provided with a simple breaker device at one or more points along its length to assist drying by disturbing or turning over the material on the band.

With very large productions where floor space is at a premium it is usual to employ multi-deck band dryers. The simplest, least expensive and generally least efficient is composed of a series of perforated bands discharging from one to the other and provided with a common air circulation and reheating system. The most efficient design and one having all the advantages of straight-through dryers is shown in Fig. 19 and Plate XIX.

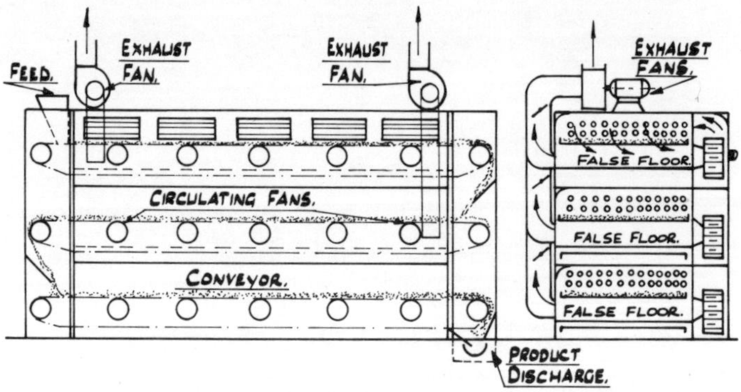

Fig. 19. Multi-deck band dryer.

In this, each deck constitutes a self-contained straight-through band dryer furnished with sections or zones having their individual recirculation fans and heating arrangements, and each dryer discharging to the next. In this design drying conditions are under maximum control and the hot exhaust air from the last deck can be fed to a previous one as is most desirable for optimum thermal efficiency. At the same time, by adjusting the band speed on each deck, the bed depth can be varied at will. For very high drying duties it is customary to feed a deep bed to the top deck, which will be self-contained as regards air supply, progressively reducing the bed depth on subsequent decks and leading the exhaust from the bottom or last deck to an earlier one. In view of the considerable evaporation on

Plate XIX. Multi-deck band dryer. *Courtesy A.P.V.-Mitchell (Dryers) Ltd.*

the top deck this is usually exhausted separately to atmosphere. Feed and discharge points on each deck will be housed and sealed off separately.

Heating Systems

The above designs of band dryer operate on the sectional or zonal re-circulation principle as described under batch forced convection ovens. The general movement of air from zone to zone can be parallel or counter to the flow of material. The principle of reheating is also employed either by suitable location of heating coils in steam-heated dryers or the points of introduction of hot gases in direct-fired dryers. Two alternative methods of applying direct-fired combustion gases are shown diagrammatically in Fig. 20, noting that in diagram (b) representing a multi-deck dryer a source of hot air at high temperature can be obtained by a similar arrangement using a heat exchanger when contact between combustion gases and the material is not permissible.

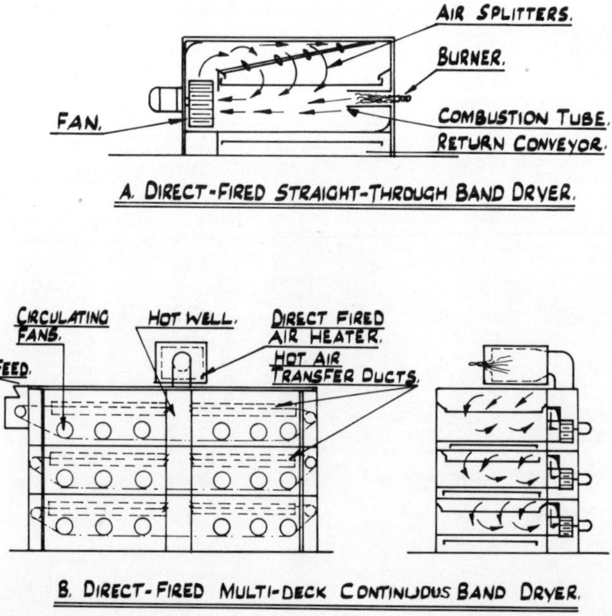

Fig. 20. Direct-heated band dryer.

In the multi-deck dryer system shown, heated air or gases are introduced to one or more "hot-wells"—depending on the length of the dryer—connected to hot gas ducts running longitudinally along the fan side compartments adjacent to the suction inlets of the fans. Louvre-operated ports at these points control the quantity of hot gases introduced into the recirculating system of each dryer section thus supplying the heat requirement of that section and maintaining it at the requisite temperature. It is usual to have an over-riding thermostatic control of the temperature of gas entering the "hot-well" and either to set individual inlet port louvres by hand or to have a mechanical linkage thermostatically controlled by the temperature in a particular section or zone of sections. This system of "hot-well" combustion gas supply is also used in single-deck band dryers of considerable length although for shorter dryers there may be an individual combustion chamber air heater located at either or both ends of the dryer serving combustion gases through longitudinal ducts with inlet ports at each section. It is usual to line such ducts both in horizontal and multi-deck dryers with a protective layer of insulating material.

The design shown in Fig. 20(a) is simple and effective and permits close control of temperature in each section. A gas or oil burner mounted on the dryer panel fires into a heat-resisting firing tube located immediately opposite the suction ring of each circulating fan. Combustion gases are intimately mixed with the recirculating air and maintain the heat requirements of each section, the temperature of the mixed gases as they approach the material on the band controlling the fuel input to the burner. Such an arrangement may initially be rather expensive due to the number of burners and controls but is highly efficient in operation.

Steam-heated band dryers rely on heat transfer from steam-heated tubes to air, the heater tubes being located either above the band or in the side chamber or as a combination of both arrangements. It is unusual to locate heater batteries below the band as any fines falling through the band may lodge on the tubes, reducing their efficiency and perhaps overheating the spilt material which, if picked up by the circulating air, may be returned to the material on the band, resulting in deterioration in quality. Heater batteries located above the band where permissible have the advantage that at fairly high steam pressures they are a source of radiant heat, thus enhancing the heat-transfer rate to the material. Air velocities in this area will, however, only be of the order of 200–250 ft/min and may not be sufficient to keep the tubes free from dust. Where dust entrainment is likely, they are best located in the fan side compartments where extremely high velocities exist.

Heaters are also installed in the side compartments rather than over the band when radiant heat is undesirable or where temporary lodgement of dust on the tubes and its subsequent deposit on the band charge is undesirable. In such cases, even if heaters are permissible over the band, it may be preferable to have plain un-gilled tubes. Batteries of heater tubes should be designed and supported in such a way that condensate is rapidly removed and so arranged that they can be withdrawn from the dryer for maintenance, repair or perhaps cleaning.

The total heater tube area in the different sections of a dryer will be such as to meet the thermal requirements of heating up and evaporating moisture during the passage of the feedstock through each section and will usually be graduated from a maximum at the wet end to a minimum towards the dry end. The drying rate curve determined by pilot-scale tests will give a good indication of the distribution of the heat load on each heater section and heater areas will be provided accordingly. Fig. 21 gives curves indicating heat-transfer rates normally experienced with gilled heater batteries, the tubes being $1\frac{1}{16}$ in o.d. and $\frac{3}{4}$ in bore and fins $2\frac{1}{8}$ in o.d. 20 s.w.g., 8 fins/in, together with pressure losses at different air velocities (superficial) and gill or fin spacing. For comparison similar curves of coefficients for finned tubes employing water at 160°F are also given.

Allowance has to be made for heating up the metal band and conveyor chains as well as the solids content of the feedstock, and this is usually allowed for in the first two or three sections of the dryer. Due

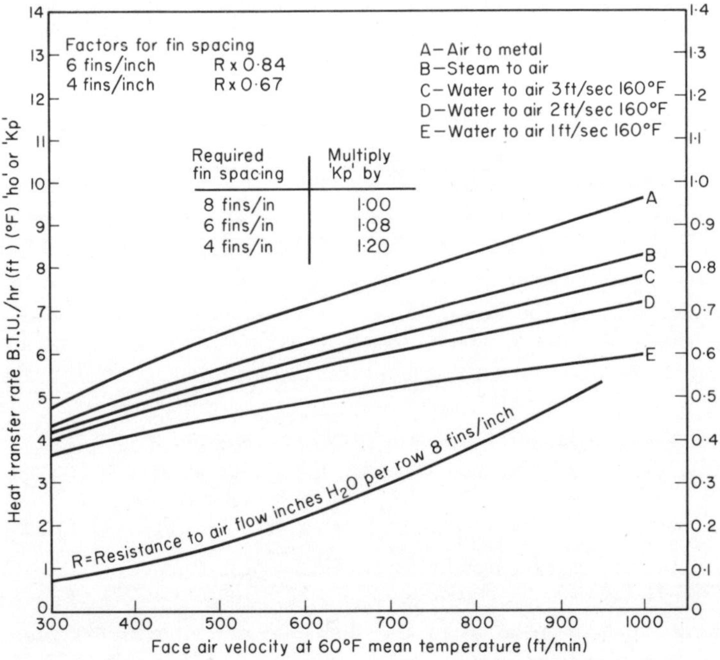

Fig. 21. Heat-transfer rates. *Courtesy Hunt Heat Exchangers Ltd.*

allowance has also to be made for heating up the cold fresh air introduced into the dryer and for heat losses by radiation and convection from the dryer external surfaces.

The quantity of fresh air introduced into a large band dryer may be considerable and, when fairly high pressure steam is employed in order to attain high drying temperatures or drying rates, a considerable economy in steam consumption can be secured by utilising condensate to pre-heat ingoing air. Increased thermal efficiencies of 10–12% can be obtained thereby. The condensate would usually have to be reduced in pressure and hence temperature before introduction to the steam-raising plant and the heat loss involved can be effectively applied in such a pre-heater. If the liquid condensate itself is not used it is usual to flash the condensate in a suitable flash vessel to a lower pressure and feed the released flash steam to the pre-heater. Fig. 22 shows the distribution of heat as between low and high pressure steam and water. The total heat content of steam does not increase greatly with increase in pressure whereas the sensible heat in condensate increases considerably with increase in pressure. As the condensate would have to be reduced below its boiling point before passing to a boiler plant, hot-well or feed-water pump, a considerable heat recovery is possible by use of an air pre-heater. An even greater heat re-

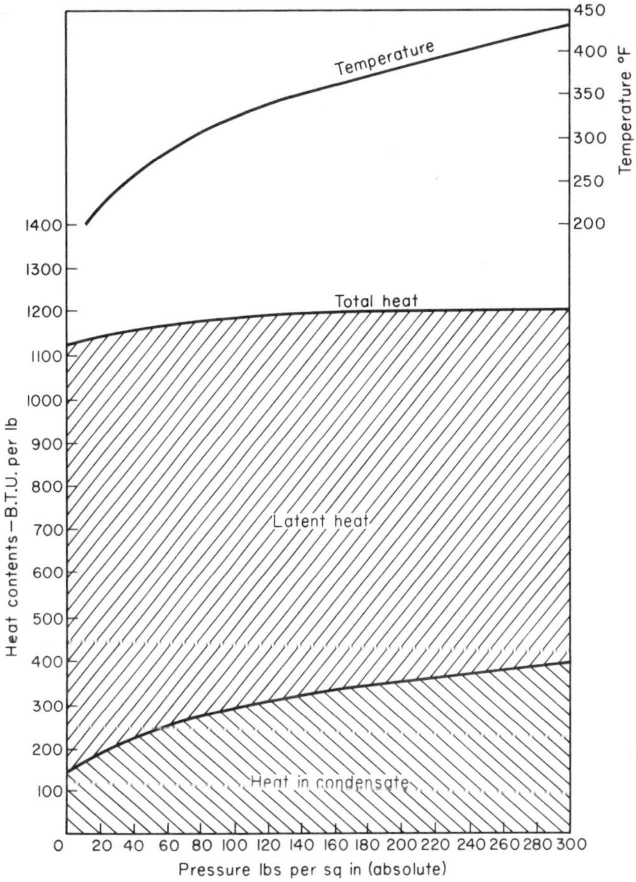

Fig. 22. Heat in steam and condensate *Courtesy H. M. Stationery Office.*

covery is possible from the condensation of flash steam at low pressures.

As the heat in condensate at a given pressure is greater than at a lower pressure, if the pressure is reduced the excess heat causes some condensate to evaporate or "flash" and thus absorb the excess heat. Fig. 23 shows the amount of steam generated when condensate at different pressures is discharged into a flash vessel operated at 10 lb/in^2. Thus condensate from a heater battery operating at 200 lb/in^2 gauge if released into a flash vessel maintained at 10 lb/in^2 gauge will form steam to the extent of 16% of the weight of the condensate; condensate at 80 lb/in^2 gauge would yield 9% flash steam. Such steam recondensed in a pre-heater to approximately atmospheric pressure would give up approximately 950 Btu/lb. As an example: suppose a band dryer has to evaporate 1,000 lb/h calling for a total heat load of 1,210,000 Btu/h and requiring 5,150 lb/h of fresh air

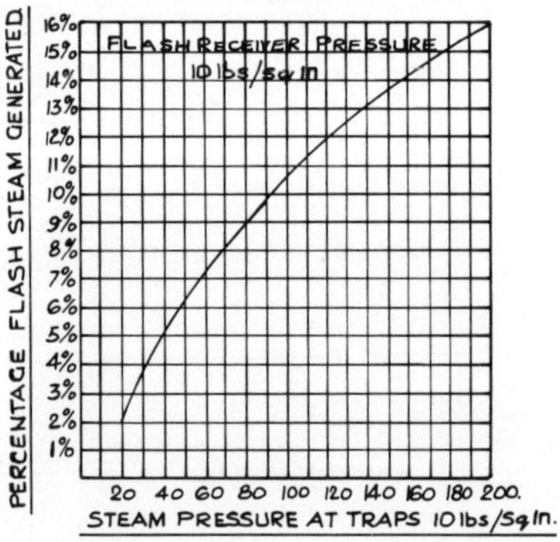

Fig. 23. Amount of flash steam generated.

to be heated from 60 to 200°F, steam being available at 100 lb/in^2 gauge. About 10% losses due to convection and radiation are assumed. Heat required to raise 5,150 lb air/hour to 200°F would be 173,000 Btu/h giving a total heat load including losses of 1,383,000 Btu/h equivalent to a steam usage of 1,694 lb/h, of which about 200 lb of steam/hour would be used in heating the fresh air. Were this fresh air to be heated by flash steam, only about 1,500 lb/h of steam would be required and this would yield 10·5% of flash steam at 10 lb/in^2 gauge namely 157 lb/h, which on recondensation in a pre-heater would make 149,500 Btu/h available. This amount of heat would be capable of raising the fresh air from 60°F to about 180°F, leaving a greatly reduced heating load for the dryer heater batteries and representing a saving of about 10·8% in thermal energy.

Sizing a Band Dryer

Although a theoretical approach to the sizing of a continuous through-circulation band dryer, based on heat and mass transfer, particle size and surface areas and driving forces can be made, the results would be too unreliable for practical purposes. The reader can only be referred to published work on this aspect for his information. Accurate data for the purpose of designing a through-circulation band dryer can only be obtained by carrying out drying tests under conditions which are appropriate for the material under investigation and can be reproduced on a production plant. Where the feedstock has to be pre-formed, the design and effect of the pre-former may have an important bearing on the drying characteristics of the material, and this aspect of the problem requires investigation along with the actual drying tests. Fin-drum pre-forming will secure a measure of pre-drying resulting in less drying in the constant rate period;

simple granulating will usually result in a material with a higher drying rate capability than pressure extrusion. The actual drying tests are quite straightforward, being a determination of the drying rates over the full drying cycle of a fixed bed area at known different depths, air velocities and air temperatures. These tests will indicate the desirable values of the different variables to yield optimum drying rates for specified bed loadings. Once the optimum drying rate is ascertained, the size of the dryer can be calculated for the required production at a specified band area loading. Commercial dryer performance may not quite equal test figures due to uneven loading from the feed arrangement and other factors for scale-up purposes. Depending upon the method of feeding this may result in an allowance of 25–50% of the calculated requirements. The total band area required will be the production of dried product per hour divided by the loading of dry product per square foot of band area multiplied by the total drying time in hours, adjusted for scale-up factor. Fig. 24 is convenient for arriving at the required effective band area assuming a 25% scale-up factor.

Thus, if a production of 1,200 lb/h dry product were required from material which at 6–8 lb/ft^2 loading of dry product takes 45 min to attain the desired residual moisture, an effective band area of about 190 ft^2 would be required assuming a 25% scale-up factor. If the band had an effective width of 4 ft 6 in the effective length of the dryer would be 42 ft and if composed of 6 ft long fan sections would probably be installed as a seven-section dryer. Really reliable values for scale-up factors can only be arrived at from extensive experience with full-scale installations correlated with laboratory tests embracing feeding and pre-forming equipment together with observation of drying tests, so that the ultimate answer will usually be provided only by the dryer manufacturer. The above method of estimating the probable size of a band dryer gives no indication of the distribution of the heat load throughout the length of the dryer, the amount of heater surface to be provided nor the amount of fresh air which will have to be introduced into the dryer and its thermal demands. The following is an example of how a drying problem would be resolved in terms of a continuous through-circulation band dryer.

PROBLEM. It is required to produce 1,000 lb/h of dry product containing 0·5% residual moisture from the cake yielded on a continuous vacuum filter having 25% by weight moisture. The filter cake is fed to an extruder at 40°F and the maximum permissible temperature of the dried product is 220°F. The bone-dry material has a specific heat of 0·2 Btu/lb and the dry product has a bulk density of 24 lb/ft^3. Steam is available at 100 lb/in^2 gauge and the ambient air in the works can be taken to be at 60°F and 55% relative humidity. Calculate the size of dryer required, the required

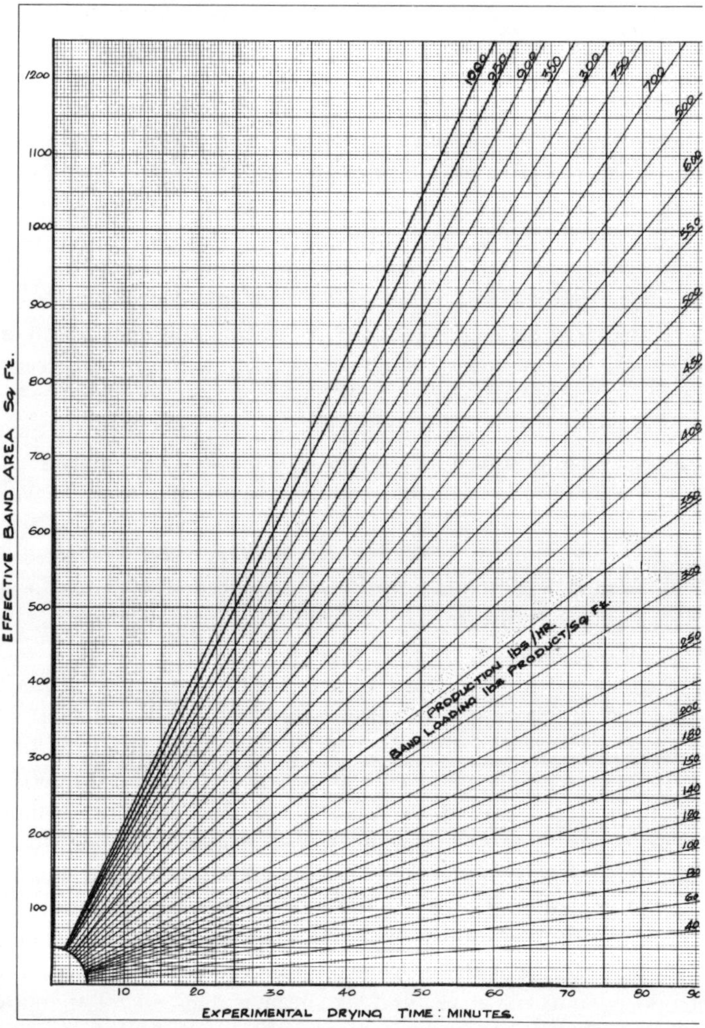

Fig. 24. Estimation of band dryer size.

heater capacity and distribution down the length of the dryer and the size of the exhaust fan, assuming no pre-heating of the fresh air and exhaust air being at 180°F and 30% relative humidity.

SOLUTION. Laboratory tests showed a roller extruder to be the most satisfactory and the drying characteristics of the extrusions on a 1 ft^2 test tray with 3 in depth of bed amounting to 8 lb of material were as follows:

Wet weight of charge	8 lb
Dry product at 0·5% moisture	6 lb

Superficial air velocity	250 ft/min
Air approach temperature	250°F
Final material temperature	205°F

Time min	*% Moisture w/w basis*	*Moisture content lb*	*Evaporation rate lb/ft²/h*
0	25	2·0	
5	22·5	1·8	2·4
10	18·5	1·48	3·84
15	15·5	1·21	3·36
20	11·5	0·92	3·36
25	7·5	0·60	3·84
30	3·5	0·28	3·84
35	1·3	0·104	2·11
40	0·5	0·04	1·2

For a production of 1,000 lb/h at 0·5 % w/w residual moisture, the evaporation will be 327 lb/h and the production of bone-dry material will be 995 lb/h. The heat requirements to dry the material with a final moisture product temperature of 205°F and an exhaust air temperature of 180°F will be:

	Btu/h
1. Heat to solids: $995 \times 0{\cdot}2 \times (205-40)$	32,800
2. Heat to water: $332 \times 1 \times (180-40)$	46,500
3. Heat to evaporate water: 327×970	317,000
	396,300
4. Add 10% for convection and radiation losses	39,630
Total	435,930

With a production rate of 1,000 lb/h and a product loading substantially $8 \times 0{\cdot}75 = 6$ lb/ft³ equivalent to 166·7 ft²/h, as the drying time is 40 min there would be 140 ft² of band within the dryer allowing a 25% scale-up factor. Assuming the dryer will have a band 6 ft wide effective and that each dryer section is 6 ft long, a four-section dryer will be required having an effective length excluding feed and delivery sections of 24 ft. Assuming the weight of conveyor chains and flights in this length of dryer to be 1,500 lb, this would require 26,000 Btu to raise it from 60°F to 205°F in 40 min or 39,000 Btu/h giving a total of $435{,}930 + 39{,}000$ or say 475,000 Btu/h.

The fresh air at 60°F and 55% relative humidity will have an absolute humidity of 0·006 lb/lb of bone-dry air (from the humidity chart) while the exhaust air at 180°F and 30% relative humidity will contain 0·114 lb/lb. The moisture picked up is therefore 0·108 lb/lb of bone-dry air fed to the dryer, the quantity of which will be 327/0·108 = 3,028 lb/h with which is associated 3,028 × 0·006 = 18·2 lb moisture, making a total of 3,046 lb/h of fresh air of specific volume of 13·5 ft^3/lb. This air will have to be heated to about 250°F by the heaters, calling for 139,000 Btu/h. The total heat requirement of the dryer will therefore be 475,000 + 139,000 = 614,000 Btu/h. Steam at 100 lb/in^2 gauge on condensing releases 898 Btu/lb so that the estimated steam requirement is 684 lb/h or 2·1 lb steam/lb water evaporated.

Turning now to the drying tests, if the percentage of moisture in the material and the evaporation rates are plotted against drying time, the following drying curves are obtained.

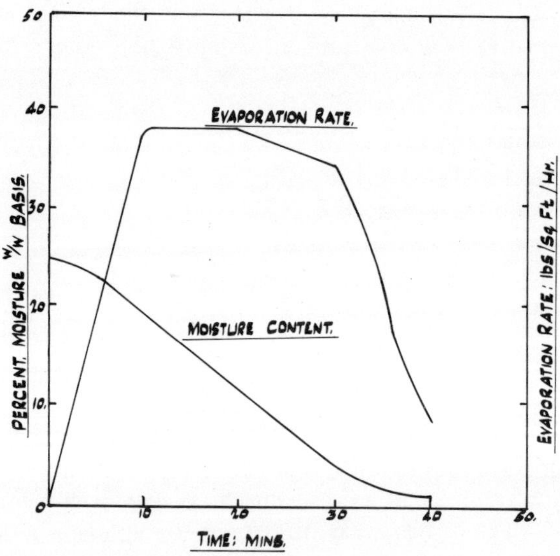

Test drying curves

Splitting down the evaporation rate-drying time curve into the equivalent of the four sections of the dryer, assuming all the heating-up of the material and band takes place in the first section and that the heat required by the fresh air is supplied by the heaters in the remaining sections, we have the following distribution of heat load.

Section	1	2	3	4
Mean evaporation rate, $lb/ft^2/h$	1·92	3·84	3·64	2·152
Evaporation, lb/h	54·2	108·8	103·1	60·8
Btu/h	52,700	105,500	100,000	59,000
Heat to conveyor, Btu/h	39,000			
Heat to water	46,500			
Heat to solids – say	16,000	16,800		
Heat to fresh air – say		50,000	50,000	50,000
Convection and radiation losses – say	10,000	10,000	10,000	10,000
Total requirement/section/Btu/h	164,200	182,300	160,000	119,000
Heater area at 900 $Btu/ft^2/h$	182·5	206	178	132·5

Assuming the radiation and convection losses equally distributed over the sections, i.e. 10,000 Btu/h per section and a heat transfer rate of 900 Btu/ft^2 of heater surface the heater areas will be:

Section 1. 182·5 ft^2; Section 2. 206 ft^2;
Section 3. 178 ft^2; Section 4. 132·5 ft^2.

Festoon Conveyor Dryer

An interesting design or dryer, although not a through-circulation band dryer in the true sense, is the screen belt festoon dryer shown in Fig. 25. This dryer is used only in conjunction with a continuous rotary filter, the wire-mesh belt being the foundation on which the filter cake is formed. The feedstock requires to be easily filterable and capable of forming a cake at least about $\frac{1}{2}$ in thick in order to fill adequately the spiral wire mesh of which the belt is composed. After leaving the filter drum, the belt is led to a festooning device and the festoons are conveyed continuously through a drying chamber in which hot air is circulated in the conventional manner. The festoon loops suspended from the conveying device remain undisturbed as they pass through the dryer, but on their arrival at the discharge end they pass over reversal rolls and are subjected to the action of a beater which removes the now dried cake from the woven belt which returns to the filter drum.

The festooning arrangement results in a considerable holding capacity of the dryer for a given floor space. With the progressive movement of the feedstock through the different sections of the dryer, the temperatures of which are thermostatically controllable, extremely uniform drying is secured and low drying times are possible without the use of ancillary pre-forming equipment. Internal steam heater coils or an external source of hot gas may be used. Typical performance figures for this design of dryer are given in Table 18.

Table 16. Band dryer performance data.

	Fertilisers	*Bentonite*	*Pigment*	*Nickel hydroxide*	*Metallic stearate*
Effective dryer length	42 ft 6 in	60 ft 0 in	24 ft 0 in	24 ft 0 in	41 ft 3 in
Effective band width	8 ft 6 in	8 ft 6 in	4 ft 0 in	4 ft 0 in	6 ft 0 in
Capacity, lb product/h	2,290	8,512	100	125	125
Method of feeding / Feedstock pre-forming	Oscillator	Oscillator	Extruder	Extruder	Extruder
Initial moisture, % w/w basis	45·0	30	58·9	75	75
Final moisture, % w/w basis	2·0	10·0	0·2	0·5	0·2
Drying time, min	16	14	60	70	60
Drying rate, lb evaporated/ft^2/h	7·0	6·5	2·0	7·5	1·5
Air temperature range, °F	—	—	—	—	—
Superficial air velocity, ft/min	200	200	180	180	125
Heat consumption, Btu/lb evaporated	—	—	—	—	—
Method of heating	Direct oil	Direct oil	Steam	Steam	Steam
Fan installed hp	35	50	14	14	28

Courtesy A.P.V.-Mitchell (Dryers) Ltd.

Table 17. Band dryer performance data

	Aluminium hydrate	*Polyacrylic nitrile*	*Sulphur*	*Calcium carbonate*	*Titanium dioxide*
Effective dryer length	32 ft 9 in	43 ft 0 in	28 ft 0 in	50 ft 0 in	108 ft 0 in
Effective band width	6 ft 6 in	6 ft 6 in	6 ft 6 in	6 ft 3 in	9 ft 6 in
Capacity, lb product/h	615	2,070	660	1,800	6,000
Method of feeding / Feedstock pre-forming	Grooved drum	Extruder	Extruder	Extruder	Extruder
Initial moisture, % w/w basis	38·0	55·0	45·0	60·0	50·0
Final moisture, % w/w basis	0·2	1·0	1·0	0·5	0·5
Drying time, min	26	52	110	40	45
Drying rate, lb evaporated/h/ft^2	2·88	3·37	3·57	5·73	6·0
Air temperature range, °F	233	186/130	194/230	320	314/392
Superficial air velocity, ft/min	140	100/216	140	160	150
Heat consumption, lb steam/lb evaporated	1·7–1·3	1·8–1·9	1·8–1·9	1·7–1·8	1·8–1·9
Method of heating	50 lb/in^2 steam	25 lb/in^2 steam	90 lb/in^2 steam	160 lb/in^2 steam	260 lb/in^2 steam
Fan installed hp (approx.)	25	65	20	35	80

Courtesy Krauss-Maffei-Imperial G.m.b.H.

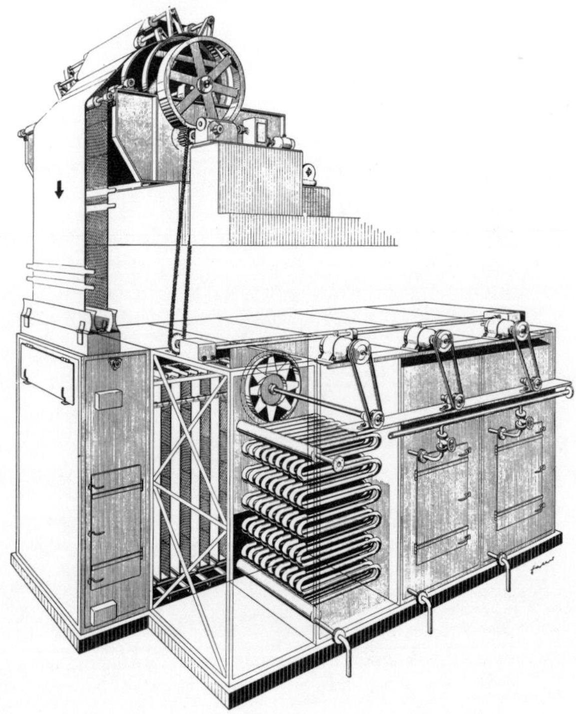

Fig. 25. The Imperial Festoon conveyor dryer. *Courtesy Krauss-Maffei-Imperial G.m.b.H.*

Table 18. Performance data for Imperial Festoon dryer

Material	*Calcium carbonate*	*Yellow iron oxides*	*White lead*	*Nickel carbonate*
Dryer height, ft approx.	15 ft 0 in	15 ft 0 in	11 ft 6 in	13 ft 0 in
width, ft approx.	12 ft 6 in	12 ft 6 in	11 ft 6 in	12 ft 0 in
length, ft approx.	35 ft 0 in	30 ft 0 in	15 ft 0 in	18 ft 0 in
Dryer surface, ft^2	6,000	3,500	520	1,100
Production, lb/h/product	1,170	1,100	555	300
Initial moisture, % w/w basis	60·0	67·0	33·0	73·0
Final moisture, % w/w basis	0·5	0·5–1·0	0·3	1·0
Drying temperature, °F	250	250/300/360	250/230	300/265
Method of heating	150 lb/in^2 (steam)	Gas	90 lb/in^2 (steam)	150 lb/in^2 (steam)
lb steam/lb evaporated	1·8–1·9	—	1·8–1·9	1·8–1·9
Installed hp	35	30	80	15

Courtesy Krauss-Maffei-Imperial G.m.b.H.

An extremely efficient and compact band dryer suitable for fairly light duties in which true counter-current drying is carried out whilst retaining the multiple re-heat principle is shown in Fig. 26.

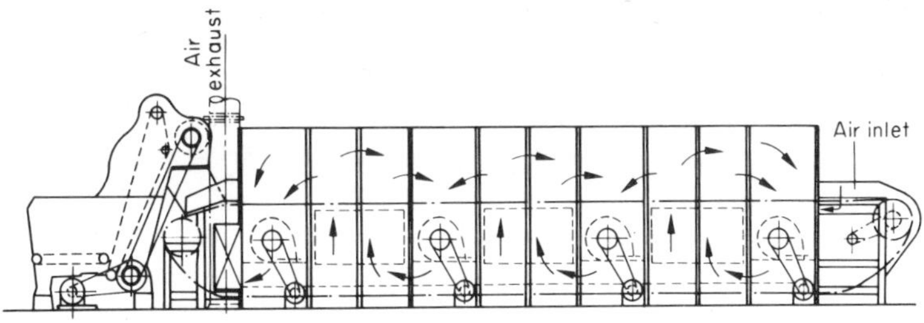

Fig. 26. Counter-current multiple reheat band dryer. *Courtesy Petrie & McNaught Ltd.*

As can be seen, counter-current air passes through the dryer alternately upwards and downwards through the bed, being picked up by centrifugal fans housed within the dryer which pass the air through a series of re-heater steam-heated units after each double passage of air through the bed. The main fresh air is introduced before the first fan but additional amounts of fresh air can be introduced on the suction side of a fan at desired points down the length of the dryer in order to maintain the exit air at a safe degree of saturation. One of the attractions of this design of dryer is that the floor space occupied by the dryer is very little more than the length and breadth of the apron as there are no side compartments for accommodating fans and heaters. Such a dryer is eminently suitable for drying textile loose stocks, as on the multiple re-heat principle drying can be accomplished at reasonably low temperatures which are amenable to close control down the length of the dryer. The width of the band will however be restricted to that demanded by uniform air distribution of one fan and heater or two fans in parallel. Furthermore as this type of dryer is usually used for fairly light loose feedstocks, superficial air velocities approaching the band are generally of the order of 150–200 ft/min as compared with the 200 to 250 ft/min of more conventional designs. This restricts the fans to more convenient size and the total volume of air exhausted.

It will be appreciated that, with conventional designs of band dryers, the feedstock being delivered to and discharged from the upper strand of the band, the return strand is empty and serves no useful purpose in the drying operation. With materials having relatively low moisture content and which are free-flowing or at no stage adhere to the band surface, it is possible by a modification in the conveyor band design to utilise both the upper and return strands. The operation of such a dryer is illustrated diagrammatically in Fig. 27. This type of dryer is often referred to as a falling-tray band dryer. In it each band is composed of a series of perforated or wire-mesh trays hinged along one side of their length and supported in a horizontal position during their passage through the main dryer section.

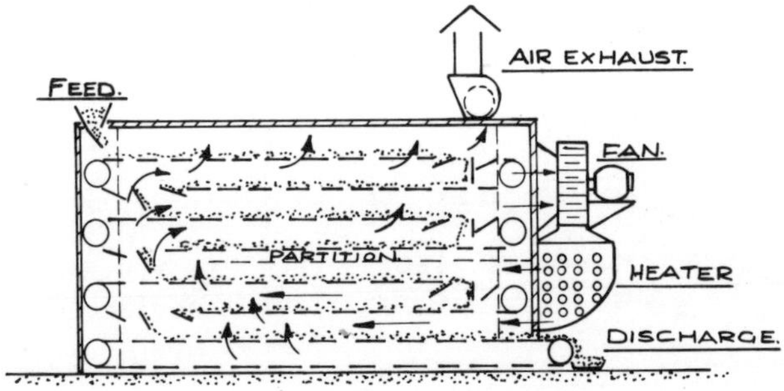

Fig. 27. Falling-tray band dryer.

Towards the end of the conveyor travel this support is omitted and the trays automatically swing into the vertical position depositing their load on to the returning strand of the same conveyor or the upper strand of a lower conveyor. In this manner both runs of the conveyor band are always fully effective so that such a dryer may have a very considerable holding capacity and can offer long residence times for a minimum of floor space. Air circulation may be horizontally along the length of the conveyor or through several conveyors in series. Such a dryer is not usually employed for high evaporational loads but is useful where a measure of conditioning of the product is required after removal of the desired amount of moisture.

Rotary Louvre Dryers

Although usually classified as a rotary dryer, the rotary Louvre Dryer relies on through-circulation for its operation and is therefore discussed in this chapter. This dryer, which is manufactured in America by the Link-Belt Company and in Great Britain by Dunford & Elliott Process Engineering Limited, has wide applications for granular and other materials which are permeable to through-circulation of air and which are reasonably free-flowing. In this dryer, feedstock is fed in at one end and discharged from the other end of a rotating shell which is not fitted with lifting flights as in a conventional rotary dryer. Instead, hot air is blown through louvres extending the full length of the dryer over the bottom section of the shell which is covered with a bed of material undergoing a rolling action from the rotation of the shell. The material is constantly exposing fresh surfaces to the hot air stream, and the shell, having the shape of an elongated truncated cone, develops an increasing depth of material as it progresses to the discharge end. Meanwhile the material may be in a more or less fluidised state depending upon the air velocity through the louvres. Fig. 28 illustrates the general principles of design and action of this type of dryer, whilst Plates XX and XXI show an industrial installation and a cut-away illustration of the internal design of the dryer.

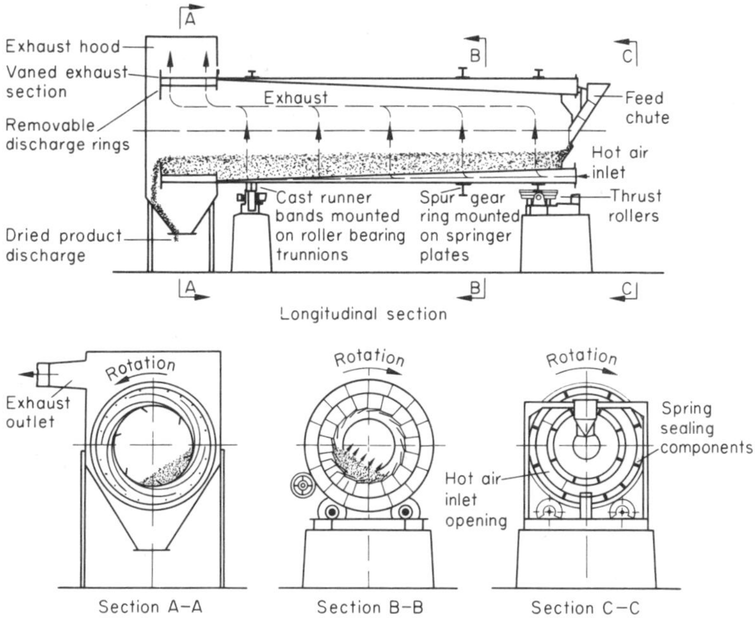

Fig. 28. Rotary louvre dryer. *Courtesy Dunford & Elliott Process Engineering Ltd*

Plate XX. Rotary louvre dryer installation. *Courtesy Dunford & Elliott Process Engineering Ltd.*

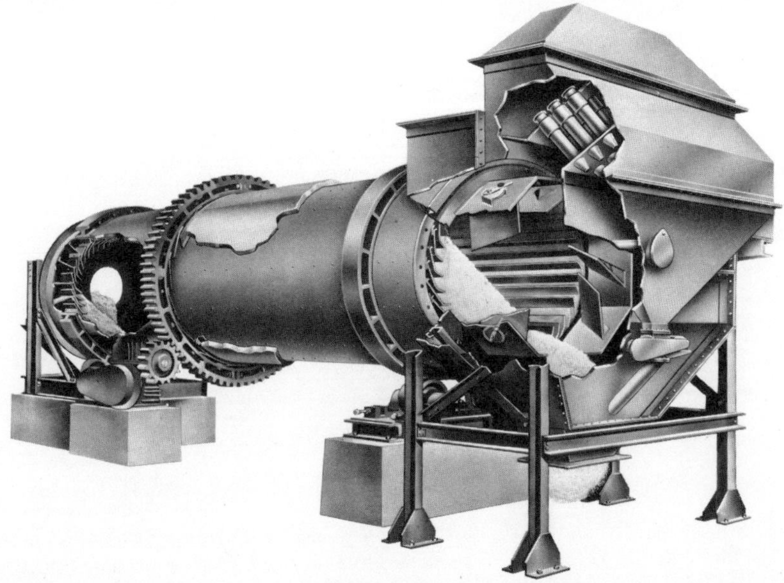

Plate XXI. Rotary louvre dryer internal design. *Courtesy Dunford & Elliott Process Engineering Ltd.*

During its passage through the dryer, the material being in a semi-fluidised state assumes a surface level which is substantially horizontal so that the bed depth gradually increases from feed end to discharge, becoming greater as drying proceeds. As the resistance of the bed to air flow is approximately directly proportional to the depth of bed, the quantity of air, and hence its velocity through the bed, will decrease towards the discharge end where the material will be drier and lighter, thus reducing the tendency to entrainment. The general effect of this difference in quantity of air passing through the bed from feed end to discharge end results substantially in a parallel flow movement of the air in the main body of the dryer whilst still retaining the optimum driving force effect of maximum permissible air temperature in the bed itself throughout the passage of material through the dryer. The air or hot combustion gas fed to the dryer is applied to the whole of the area of that segment of the dryer embraced by the gas feed duct and the louvres presented to it, which is the length of the dryer multiplied by about 65% of the dryer diameter so that the superficial velocity of the gases through the exposed louvres will be much lower than if the same quantity were fed to a conventional rotary dryer. Furthermore, since the gases are released into the main body of the dryer shell progressively down its length, their velocity in the shell will be low, gradually increasing towards the exit end but not excessively so in view of the gradually increasing shell diameter. Entrainment danger in both material bed and in the shell is therefore minimised whilst retaining optimum velocities within the material bed.

The rotation of the dryer promotes intimate mixing and turning over of the feedstock, with the result that fresh surfaces are constantly exposed to the incoming hot gases, resulting in very uniform drying. Similarly fresh louvre surfaces are arriving at and leaving the area exposed to the highest temperatures, thus restricting their rise in temperature and in many cases obviating the use of heat-resisting alloys in their construction. Due to the design of the louvres, very little relative movement occurs between louvre surface and feedstock, thus reducing attrition of the latter and abrasion of the former.

A further advantage of this dryer is that it can be operated either as a constant temperature or as a graduated temperature continuous rotary dryer, which is virtually impossible with conventional rotary dryers. The driving force for drying can therefore be maintained constant throughout the length of the dryer in terms of temperature or may be varied stepwise to suit the material undergoing drying depending on whether a single or multiple gas feed is used at one or more temperature levels. This includes, of course, the possibility of employing part of the dryer length as a cooling zone.

A wide variety of products is successfully handled in this design of dryer as will be seen from the following typical selection:

Material	*Moisture reduction wet basis percentage*	*Inlet air temperature °C*
Aspirin	5 to 0·1	50
Bone char	32 to 3·0	250
Bone meal	37 to 3·0	325
Calcium citrate	50 to 0·2	300
Carbon black	40 to 0·0	450
Cellulose acetate	72·9 to 2·0	165
Clay (china)	27 to 5·0	400
Clay (ball)	18 to 4·0	400
Coal	20 to 2·0	400
Diatomaceous earth	72 to 30·0	450
Dicalcium phosphate	50 to 0·0	300
Methyl cellulose	72 to 1·3	160
Polystyrene	20·8 to 1·42	140
Pyrites (flotation)	7 to 0·5	350
Salt (NaCl)	7 to 0·013	170
Sand	20 to 0·1	500
Sand	8 to 0·1	500
Sugar	1·5 to 0·03	90
Sulphate of ammonia (20°C)	2·5 to 0·10	140
Sulphate of ammonia (50–70°C)	2·5 to 0·10	140

Table 19 gives industrial plant operating performance data for five different products.

Table 19. Performance of rotary louvre dryer

	Bone meal	*Sugar**	*Sulphate of ammonia**	*Bread crumbs*	*Bentonite*
Dryer diameter	7 ft 6 in	7 ft 6 in	7 ft 6 in	4 ft 6 in	8 ft 10 in
Dryer length	12 ft 0 in	25 ft 0 in	25 ft 0 in	20 ft 0 in	30 ft 0 in
Initial moisture, % w/w basis	17·0	1·5	1·0	37	45
Final moisture, % w/w basis	7·0	0·03	0·2	2·5	11
Method of feed	Screw	Screw	Chute	Chute	Chute
Evaporation rate, lb/h	1,660	500	400	920	7,100
Efficiency Btu supplied/lb evaporation	74·3	40	—	55	62·5
Method of heating	Steam	Steam	Steam	Gas	Oil
Inlet air temperature, °F	203	194	248	572	842
Outlet air temperature, °F	122	104	149	158	176
Residence time, min	9·3	12·5	9·0	25·7	37·3
Fan hp (absorbed)	49·3	52·2	55	13·7	54·3
Motive hp (absorbed)	8	12·5	15	2·3	20·0
Fan capacity, ft^3/min					
Inlet	9,560	18,000	16,000	5,380	20,000
Outlet	14,000	22,300	21,000	5,100	25,000

Courtesy Dunford & Elliott Process Engineering Ltd.

*Combined two-stage dryer–cooler.

CHAPTER VII

Rotary Dryers

With the possible exception of tray drying ovens, rotary dryers are more widely used in industry than any other type. This is understandable in view of the very wide range of materials which can be dried in them and because if the correct design is selected for a particular application and is correctly operated quite high thermal efficiencies are possible. They require little or no labour to operate and if properly maintained, particularly as regards lubrication, they can be operated continuously over lengthy periods under automatic control with occasional supervision.

A rotary dryer operated at atmospheric pressure consists of a cylindrical shell, ranging in length from four to ten times its diameter, into which the wet charge is fed continuously at one end and is discharged at the other, the movement of the material being due usually to the combined effect of inclination of the shell to the horizontal and the action of internal lifting flights during the rotation of the shell. The latter is usually provided with machined girth rings supported on suitably designed bearings. Rotation of the shell is secured by the action of suitable gearing or, in some cases, a chain drive.

Direct Rotary Dryers

When the materials to be dried can be safely brought into contact with heated air or combustion gases and are not too dusty, either originally or as a result of drying, and when volatile inflammable or noxious components are absent except in small quantities, direct rotary dryers are used. The heating and drying medium is introduced into the dryer and may flow either parallel with or counter to the flow of material. When flowing parallel with the charge, the drying medium will, to a greater or lesser extent, assist the flow of material through the dryer. In some designs flow may be both parallel and counter, and in both indirect and direct contact with the material. This type is generally referred to as an indirect-direct rotary dryer. A further type, typified by the Roto-Louvre Dryer, although a rotary dryer, is designed as a through-circulation dryer, the drying medium being passed through a permeable bed of material passing through the dryer shell, and has therefore been described in Chapter VI.

A typical direct rotary dryer, in this case a parallel-flow dryer, is shown broken down into component sections in Fig. 29. It will be seen that it

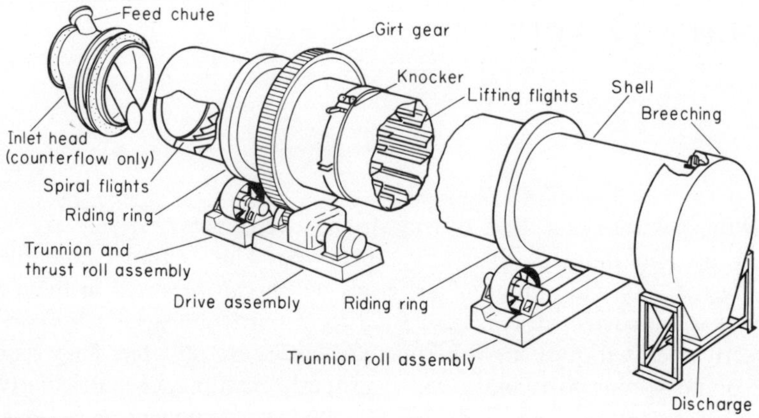

Fig. 29. Direct single shell rotary dryer.

consists essentially of a connection for the introduction of the heating and drying medium, a feed connection for wet material, the main cylinder or shell provided with machined girth rings riding on suitable bearings, a girth gear ring, pinion and reduction gear drive, a discharge hood from which dry product and humid gases are discharged and internal flights. Flights at the material feed end will usually be spirally arranged to propel the feedstock fairly rapidly from the feed point into the dryer, followed by additional flights running longitudinally down the length of the dryer. There may be a second hood at the feed end to accommodate the heat source inlet connection and the rotating shell. Suitable seals, both dust- and air-tight as far as possible, are provided between the stationary sections and the rotating shell. A counter-current direct rotary dryer will be constructed on similar lines except that feedstock entry and hot gas entry will be at the opposite ends. The gas exit hood will be provided with a gas-exhausting duct connected to such dust recovery equipment as may be required and, as this type of dryer is best operated under a mild suction, to an exhausting fan. Where the dryer is operated under balanced pressure throughout, a fan at both inlet and exhaust ends is used.

Typical arrangements of direct parallel-flow, direct counter-flow and

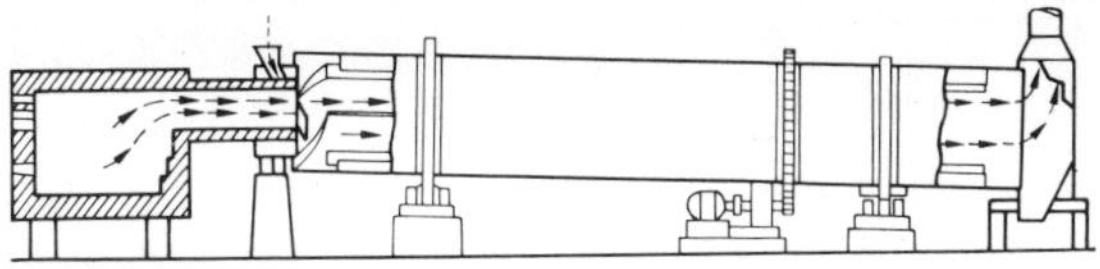

(a) Direct parallel-flow rotary dryer.

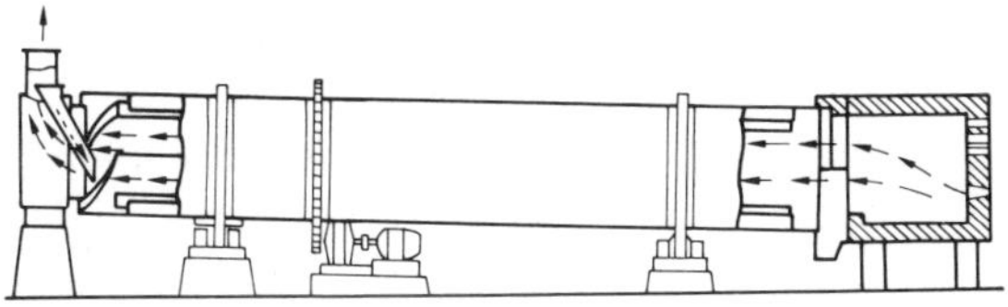

(b) Direct counter-flow rotary dryer.

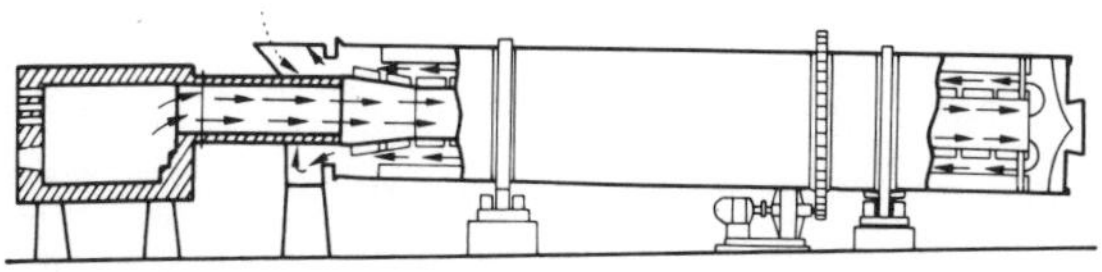

(c) Double shell indirect-direct rotary dryer.

Fig. 30. Direct and indirect rotary dryer.

double shell indirect-direct rotary dryers are shown in Fig. 30. A typical rotary dryer installation is shown in Plate XXII.

Feeding arrangements for the wet feedstock will take different forms depending largely on the physical form in which the material is presented to the dryer. Whatever form is used it should ensure as far as possible a uniform feed rate so as to disturb the operating conditions as little as possible. Further it should be such that in use it creates to the greatest possible extent a sealing effect so that cold air does not enter and hot gases do not escape from the dryer. Free-flowing feedstock will usually enter by means of an inclined chute fed from a rotary valve, a vibrating feeder or rotary table or a belt conveyor, the chute having a counter-balanced flap valve to promote a seal. Pasty and other excessively wet material may be fed by a screw conveyor or even a suitably designed pump. In parallel-flow dryers operating with high inlet gas temperatures, the chute may be water-cooled if the material is heat-sensitive. In the case of very wet materials which present feeding problems it is not unusual to back-mix some of the hot dried product with the feedstock and so reduce the latter to a feedable condition. This may be accomplished either through the use of mechanical conveyors or, in the case of considerable carry-over to a dust cyclone, by locating the cyclone above the back-mixer and introducing the hot recovered dust into the back-mixer.

Plate XXII. Rotary dryer installation. *Courtesy Buell Ltd.*

Discharge of dried product is secured either through a counter-balanced flap valve or rotary valve or a screw conveyor located at the base of the discharge hood. Where there is a high proportion of fines in the material it is possible in special instances to entrain them deliberately in the drying medium and recover them separately from a high efficiency cyclone, the coarser material being allowed a longer residence time within the dryer. The two grades of product may be collected separately or blended in the proportion in which they are produced as is most desirable.

The sealing arrangements between stationary and rotating parts of the dryer will depend upon the fineness and flow characteristics of the wet feed and dry product and the degree of positive or negative pressure at which the dryer is operated. In general it is desirable to operate the dryer under balanced or slight suction conditions so as to reduce the leakage of hot gases and of dust out of the dryer. Sealing arrangements may be simple or elaborate and their cost will vary accordingly. A variety of designs is shown in Fig. 31, but many other designs are of course in use. Whichever type is used, it is important that the dryer shell should have an absolute minimum of eccentricity in its construction. The importance of sealing the material feed arrangements has already been referred to. Where

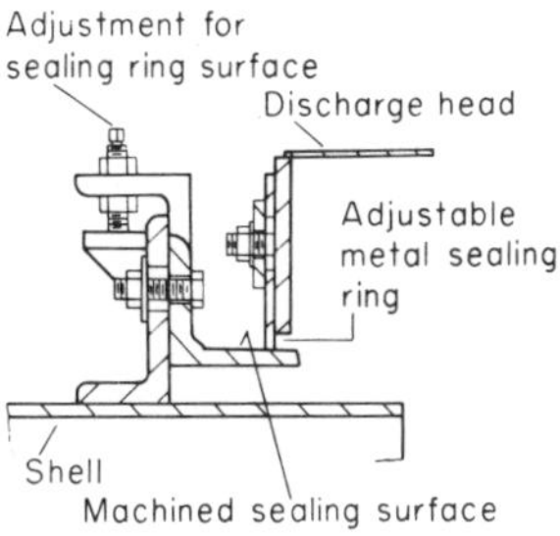

(a). Discharge end clearance-type seal assembly.

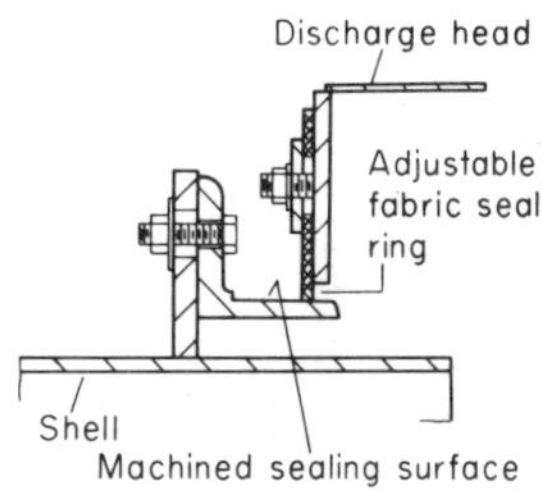

(b) Rubbing-type seal assembly.

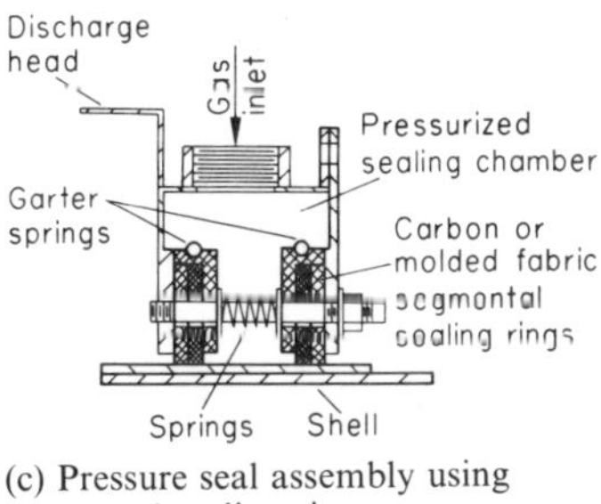

(c) Pressure seal assembly using segmental sealing rings.

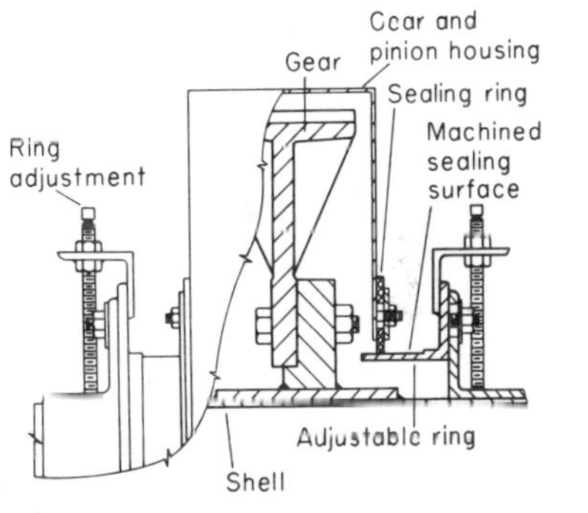

(d) Details of the sealing arrangement on a totally enclosed Ruggles-Coles gear housing.

Fig. 31. Typical rotary dryer sealing arrangements.

Courtesy Hardinge & Co. Metal Products Division of Koppers Co. Inc.

the nature of the feedstock is such as to make effective sealing impossible the dryer should be operated at, or at least as near as possible to, atmospheric pressure at the feed end, since serious air leakage at this point will nullify the effect of efficient seals between the hoods and the rotating shell.

The most effective design of rotary dryer shell seals employs heat- and wear-resisting friction material, such as brake-lining material riveted to a continuous or two-section steel ring and spring-loaded to maintain a close rubbing contact with a true-faced ring welded to the stationary hood, but the absence of eccentricity of the shell itself is all-important.

The shell of the dryer is fabricated of such material as the chemical properties of the feedstock and the temperature of operation dictate. For most uses mild steel is used but, where there is danger of corrosion or contamination of the product, stainless steel or mild steel clad with stainless steel internally is frequently used. Occasionally, for reasonably low-temperature operation, aluminium shells are used of ample section. For mild protection against corrosion or contamination, a mild steel shell may be metal-sprayed. Girth rings and supporting roller bearings should be of ample face width and, in the case of bearings, diameter, and should preferably be machined cast steel or semi-steel. The rollers should be supported on spindles of ample diameter by means of self-aligning roller bearings. The complete bearing assembly housing should be furnished with adjusting screws so that any tendency for the dryer shell to move longitudinally can be corrected. It is usual to fit a thrust roller bearing at one of the girth rings—usually that nearest the drive assembly—to take care of longitudinal movement of the shell. The girth ring will be carried from the shell by means of steel chairs suitably spaced around the shell circumference. The girth or ring gear should, whenever possible, be machined semi-steel to give smooth rotation to the shell with minimum power losses. Small diameter shells, or those of light construction, may be fitted with sprocket and precision chain drive.

Flights

The effectiveness of a direct rotary dryer, apart from the quantity and temperature of the ingoing heating medium, depends upon the extent and uniformity of the feedstock contact with the hot gases and the residence time of the material in the dryer. These will largely be a function of the number, size and design of the lifting flights. During the rotation of the shell the lifting flights pick up material and shower it as a curtain in the path of the hot gases. The depth and design of the flights will determine whether this curtain extends across the full area of the cross-section of the dryer, and the holding capacity of the flights in their travel, together with any layer of material in the lower regions of the dryer down its length, will determine the hold-up in the dryer and hence its residence time. The actual shape of the flights will, of course, to some extent depend on the physical nature of the feedstock during the course of drying. For very sticky or very wet feedstock the flights will be of a straight radial design without any lip for the initial length of the dryer, a lip to an increasing extent being introduced along the length of the dryer as drying proceeds and the product becomes more free-flowing. For free-flowing feedstocks flights with single or even double lips are used, these two forms being possibly alternated. Flights may be installed as continuous lengths down the length of the dryer or may be staggered with a view to improving showering and distribution. For very large diameter dryers, internal partitions may be introduced to give a cruciform or cellular cross-section, each compartment being provided with its own lifting flights. The performance of such an arrangement is excellent but raises problems in connection with internal cleaning, maintenance and repairs. With straight-through or staggered flights, good practice indicates that for most commercial installations their number should be between two and four times the diameter measured in feet, while the depth of flight should be between $\frac{1}{12}$ to $\frac{1}{8}$ of the diameter.

Hold-up in the dryer will depend upon the design and number of flights used, the speed of rotation and the angle of inclination of the dryer and will be arranged to suit the drying characteristics of the material. It will usually lie between 3–12% of the volume of the dryer, 7–8% being a very usual figure. As the residence time in the dryer will be the hold-up divided by the throughput, slow-drying materials and materials having a lengthy constant rate period call for adequate hold-up on flights. Hold-up on the dryer bottom in a manner experienced in a continuous kiln is to be avoided, and a low overall hold-up usually results in erratic performance difficult to control. Hold-up and residence times in parallel-flow rotary dryers tend to be less than in counter-current dryers and with some materials this effect may be such as to require a negative slope of the dryer in order to attain the requisite residence time. Where the material is liable to suffer excessive attrition through the showering effect of the flights in a parallel-flow dryer, the flights are frequently omitted for the last few feet to reduce the tendency to dust entrainment. Fig. 32 illustrates in diagrammatic form different flight arrangements.

Residence Time

Progress of material through a direct rotary dryer is affected by a number of factors,[3] namely:

Percentage loading within the dryer;
Number of flights distributed around the circumference;
Design of the flights;
Slope of the dryer from the horizontal;
Speed of rotation of the dryer shell;
Length of the dryer effectively used;
Diameter of the dryer;
Physical properties of the material;
Air velocity within the dryer.

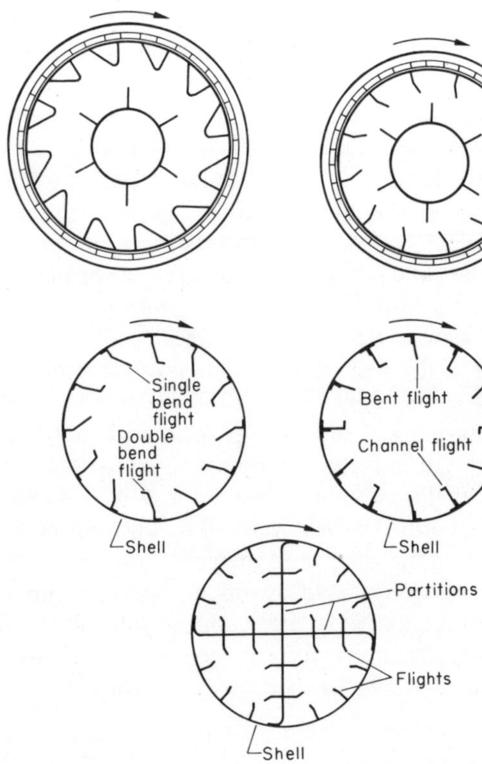

Fig. 32. Rotary dryer flight arrangements.
Courtesy Hardinge & Co. Metal Products Division of Koppers Co. Inc.

The effect of these variables in the design of the dryer on the retention time of the material in the dryer may be summarised as follows:

(1) *Percentage loading*, that is, the ratio of hold-up within the dryer to the dryer volume. Unflighted material within the dryer becomes subject to a rolling action similar to that which occurs in kilns and is to be avoided as it reduces the retention time. Further, as shown in Fig. 33, there is an optimum flight loading beyond which this "kiln" effect asserts itself and retention time decreases. Depending upon the slope of the dryer, this loading usually lies between 8–12% of the dryer volume.

(2) *Number of flights*. For a given design of flight an increase in the number of flights around the circumference of the shell will tend to have two opposing effects. The less number of times the material is lifted by the flights, the longer the retention time. This, in turn, increases the tendency to "kiln" action so that increasing the number of flights will reduce the "kiln" action and hence improve the residence time. In practice it is found

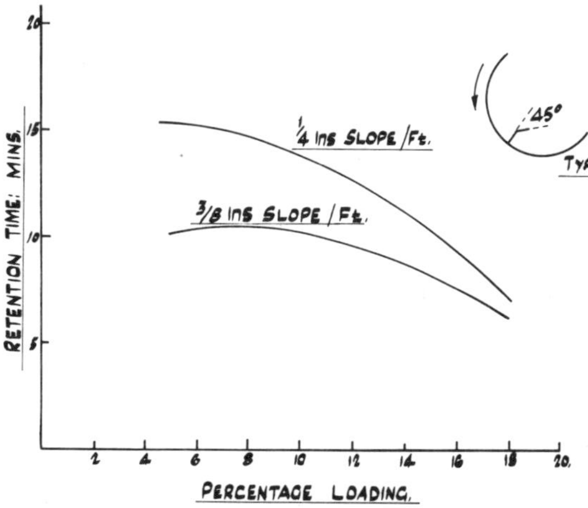

Fig. 33. Effect of per cent. Loading on retention time. *(after Brutton, Miller & Schnette).*

that decreasing the number of flights has the net result of decreasing residence time at optimum loading conditions and for a given slope and speed of rotation of the dryer. It will be appreciated that decreasing the number of flights will also decrease the exposed area of material and hence reduce the drying capacity of the dryer.

(3) *Flight design.* As the effectiveness of a direct rotary dryer will depend to an important extent on producing as uniform a curtain of material as possible across the full area of the dryer all the way down its effective length, the shape of the flights will be an important factor. The height of the flight and the extent and manner in which it is lipped will determine its maximum holding capacity and the area of exposed surface. The number and distribution of flights of differing designs will determine the extent to which the full cross-sectional area of the shell will be covered by a falling curtain of material, particularly with free-flowing material or after such a free-flowing condition is attained in the course of drying. It is frequently advantageous to employ sections of flights staggered one with the other, but these sections should not be too short so as to avoid gravitational movement of the free-flowing charge along the flight lengths with its subsequent discharge locally at open ends. Where the charge is initially pasty or very wet, it is advisable to use a smaller number of simple radial flights adequately spaced apart to avoid permanent hold-up of material.

(4) *Slope of shell.* The effect of the slope of the dryer shell cannot be considered separately from the effect of speed of rotation as each has a bearing on the other. As will be seen from Fig. 34, retention time at a given speed

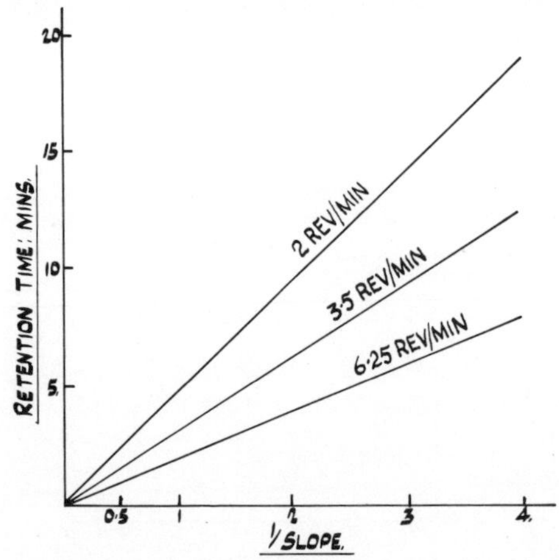

Fig. 34. Effect of dryer shell slope on retention time. *(after Brutton, Miller & Schnette).*

of rotation of shell is inversely proportional to the slope of the shell over the ranges of slope and speeds generally met with in practice, namely $\frac{1}{4}$–$\frac{3}{4}$ in/ft and 2–7 rev/min, respectively. However, at very slow speeds of rotation there is a tendency towards gradual build-up of material towards the feed end of the dryer which has the effect of artificially increasing the slope and hence decreasing the residence time.

(5) *Rate of rotation.* It is found generally in practice that, when flights of any design are at their optimum loading, the retention time is inversely proportional to the speed of rotation irrespective of flight design and spacing, but as mentioned earlier very slow speeds are conducive to kiln action which tends to reduce retention time. Rotational speed (r.p.m.) $\times$ dryer diameter (ft) usually lies between 25 and 35.

(6) *Length of dryer.* The greater the length of the effective lifting flights, the greater the retention time. It is important therefore not to have too great a length of spiral or helical flights at the feed end and, where flights are omitted altogether at the discharge end as in a parallel flow dryer, to reduce this length of the dryer to a minimum compatible with dust settlement.

(7) *Diameter.* For a given slope of shell and air velocity, the height through which the charge falls from the lifting flights will directly affect the forward movement of material.

Fig. 35 illustrates this effect; if the slope of the dryer is two degrees, and neglecting the effect of air velocity, the angle of fall of material will also be two degrees from an imaginary shell diameter. As the diameter of the shell is increased the advance of the material within the dryer will correspondingly increase.

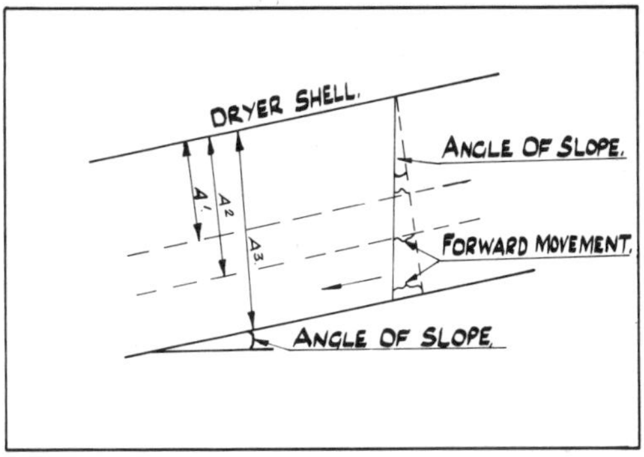

Fig. 35. Effect of Dryer Shell Diameter on forward movement of material. *(after Brutton, Miller & Schnette)*

(8) *Physical properties of charge.* Most materials other than those which are initially free-flowing will, of course, undergo considerable change in physical characteristics as drying progresses. In the initial stages, when they are still fairly wet, they will tend to accumulate at the wet end resulting in a relatively slow motion down the dryer. The rate of movement will tend to increase as drying proceeds and as break-down of lumps takes place, leading finally to a free-flowing condition. The angle of repose of free-flowing materials does not appear to have an important bearing on residence time no matter what the design of flight nor the slope of dryer within the usual range met with in practice.

(9) *Air velocity.* Air velocity has an important bearing on residence times both in parallel-flow and counter-flow rotary dryers. With light materials and materials which are dusty or become dusty by attrition through the cascading effect of the flights, air velocities must be kept at a figure where carry-over does not present a serious problem, resulting in such low residence times as to produce inadequately dried material, and this is particularly so in parallel-flow dryers. Limiting air velocities from this standpoint will therefore depend on the physical characteristics of the material. Ignoring this effect, however, the influence of air velocities can be greater in counter-current dryers than in parallel-flow dryers, as can be seen from Fig. 36. In practice air velocities range from 0·75 ft/s for fine materials to 35 ft/s for coarse heavy materials. Suitable air velocities must be determined on a pilot-scale unit operating in the manner intended on the commercial scale.

Increase in air velocities is found to increase residence time at a greater rate in counter-flow rotary dryers than the rate of reduction in residence time in parallel-flow dryers. Retention times plotted against air velocity appear to follow a straight line in the case of parallel-flow whereas the same plot for counter-flow shows a distinct curvature above a certain critical

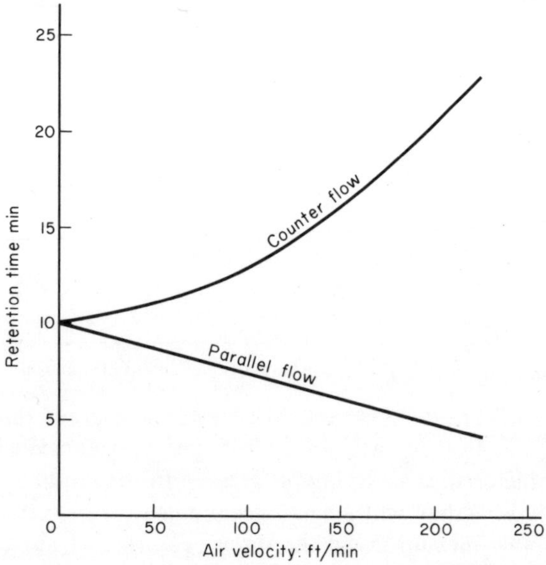

Fig. 36. Effect of air velocities on residence times.

air velocity which will depend upon the ratio of screen size to particle density.

Experimental work has resulted in arriving at empirical equations from which residence times can be approximately calculated taking into account the effect of the above-mentioned variables. Brutton, Miller and Schnette[3] give the following expressions:

$$T = \frac{KL}{nDS} + Yv$$

where T = retention time, min

L = effective length of dryer, ft

n = rev/min

D = diameter of shell, ft

S = slope, in/ft

v = air velocity, ft/min

K and Y = constants

Constant K is a function of the characteristics of the dryer design depending on the number of flights and flight design, whilst constant Y depends on the ratio of material screen size to particle density, and the method of operating the dryer, whether parallel-flow or counter-flow.

B. A. Smith[4] arrives at a simpler equation, namely:

$$T = \frac{6KL}{nDS} \text{ for single-shell direct dryers}$$

and $$T = \frac{9KL}{nDS} \text{ for double-shell direct dryers}$$

assuming 10–15 per cent flight hold-up. He gives a value of K of 0·5–2·0 for counter-flow dryers and 0·2–0·17 for parallel-flow dryers.

The drying of a material in a direct rotary dryer can be considered to take place in three stages.

(1) Heating the wet material and its moisture content to the constant rate drying temperature, which will be approximately that of the wet-bulb temperature of the drying medium.
(2) Drying of the material substantially at this temperature.
(3) Heating up of the material to its discharge temperature and evaporation of the moisture remaining at the end of stage (2).

Each stage entails its own heat-transfer effect with individual material temperature increases and heating medium temperature reductions. Treated as an overall heat-transfer operation it can be covered by the expression

$$Q_t = U_a V(\Delta t)_m$$

ignoring radiation and condensation effects, where

Q_t = total heat transferred, Btu/h
U_a = volumetric heat transfer coefficient, Btu/h/ft³ dryer volume
V = dryer volume, ft³
$(\Delta t)_m$ = overall mean temperature difference between material and gases, °F.

Determination of the temperature difference between material and gases at each end of the three stages of drying is difficult, so that an overall treatment of the heat-transfer mechanism covered by the above expression is more practicable, and a reasonable appreciation based on inlet and outlet temperatures is normally used. Furthermore, where the feedstock is high in moisture content $(\Delta t)_m$ can be considered to approximate to the logarithmic mean between the wet-bulb depression of the inlet and outlet gases.

Direct-heated Rotary Dryer Sizing

Extensive work on pilot-scale units by Friedman and Marshall[5] resulted in arriving at a relationship between the mass velocity of the gases and dryer diameters from which U_a can be evaluated with a reasonable degree of accuracy for design purposes and is given by the expression

$$U_a = \frac{10G^{0\cdot16}}{D}$$

where G = air mass velocity, lb/h/ft^2 dryer cross-sectional area
D = dryer diameter, ft.

An acceptable value for G will of course depend on the nature of the material undergoing drying and will be determined by pilot-scale tests unless a conservative figure is selected based on experience with reasonably similar materials. In practice, permissible air mass velocities can vary over a considerable range from 200 to 10,000 lb/h/ft^2.

The heat load Q_t of the dryer operation can be easily calculated assuming reasonable permissible material discharge temperatures. Air or gas inlet temperature can be assumed at a practical figure based on consideration of the material undergoing drying and the method of heating. The exit gas temperature, however, requires to be estimated in order to arrive at a value of $(\Delta t)_m$, and this is best done on the basis of calculating the number of heat-transfer units. The relationship between heat-transfer coefficient and heat-transfer unit is given by[(6)]

$$L_t = \frac{Gc^p}{U_a}$$

where L_t = length of one heat-transfer unit, ft
c_p = specific heat at constant pressure of the heating medium, Btu/lb/deg F

Substituting for $U_a = \dfrac{10G^{0\cdot16}}{D}$

we have $L_t = 0{\cdot}1\, c_p G D^{0.84}$

The number of heat-transfer units N_t will be the difference between the inlet and outlet gas temperatures divided by the overall mean temperature difference between the gases and material in °F, i.e.

$$N_t = \frac{t_1 - t_2}{(\Delta t)_m}$$

and it has been found in practice that the value of N_t should lie between 1·5 and 2 for efficient operation.

Where the moisture content of the feedstock is high and the greater part of the heat requirement is for evaporating water, the mean temperature difference can be assumed to be the logarithmic mean average of the wet-bulb depression of the inlet and exit air or gases, when

$$N_t = \log_n \frac{t_1 - t_w}{t_2 - t_w} = 2{\cdot}3 \log_{10} \frac{t_1 - t_w}{t_2 - t_w}$$

t_1 and t_2 being the dry-bulb temperature of the inlet and exit gases and t_w the wet-bulb temperature of the ingoing gas, all in °F.

In practice, best results with direct-heated rotary dryers are found feasible when the ratio of length of shell to diameter lies between 4 and 10, the more usual range being 6 to 8.

Thermal efficiencies usually experienced with direct rotary dryers vary widely, depending upon the inlet and exit air temperatures, the initial and final moisture contents of the product and whether they are counter-flow or parallel-flow dryers. Steam-heated dryers have efficiencies ranging from 30–55%, whereas high inlet temperature fuel-fired dryers will range from 50–75%. Evaporation capacities will usually be 0·15–2 lb/ft³ of dryer volume for steam-heated hot air dryers; corresponding figures for fuel-fired direct dryers will range from 1·8–7·8 lb/ft³ dryer volume. Highest values will usually occur with parallel-flow dryers fed with high moisture content stock drying down to 5–15% final moisture content.

The following is an example of the method of arriving at a suitably sized direct rotary dryer to suit a specified duty.

It is desired to produce 1,000 lb per hour of dry fine crystalline material from a feedstock containing 25% of moisture on the wet-weight basis, the product to contain not more than 0·1% residual moisture. All of the moisture is surface moisture. Tests on a pilot-scale rotary dryer indicated that a mass air velocity of 750 lb/ft²/h was permissible without undue dust carry-over. The maximum permitted temperature of the dry crystals is 275°F. Steam is available at 150 lb/in² gauge. The dry material has a specific heat of 0·22 Btu/lb and a bulk density of 50 lb/ft³. The wet crystals will be delivered to the dryer at 60°F and the ambient air in the dryer house can be taken to be 60°F with a humidity of 0·008 lb/lb of dry air.

As the material has to be dried to such a low moisture content and is heat-sensitive, a counter-current steam-heated direct rotary dryer will be selected. Steam at 150 lb/in² gauge has a temperature of 358°F and it would be reasonable to assume that this could produce a feed air temperature of 290°F. The total evaporation will be 333 lb/h. Air initially at 60°F and $H = 0{\cdot}008$ when heated to 290°F without increase in humidity will have a dew point of 104°F. Assume that the rotary dryer will consist of one and three-quarter transfer units, i.e. $N_t = 1{\cdot}75$.

$$N_t = \log_n \frac{(290-104)}{t_2-104} = 1{\cdot}75$$

whence $t_2 = 136°\text{F}$

The heat duty on the dryer will be:

		Btu/h
Heat to raise dry material	$1{,}000 \times 0{\cdot}22 \times 200$	44,000
Heat to heat moisture to 104°	$333 \times (104-60)$	14,680
Heat to evaporate moisture	$333 \times 1{,}035$	345,000
Heat to raise moisture from 104° to 136°	$333 \times 0{\cdot}446 \times 32$	4,750
		408,430

Although the dryer will be well insulated it is as well to allow about $7\frac{1}{2}\%$ for heat losses by radiation and convection, bringing the total thermal duty to 439,000 Btu/h.

This amount of heat will be liberated by the hot air fed to the dryer in cooling from 290°F to the exit temperature of 136°F and the air requirement will therefore be

$$\frac{439{,}000}{0{\cdot}24 \times (290-136)} = 11{,}850 \text{ lb/h}$$

As tests showed that a mass air velocity of 750 lb/ft^2/h was permissible, the diameter of the dryer will be

$$D^2 = \frac{11{,}850}{0{\cdot}785 \times 750}$$

whence $D = 4{\cdot}48$ ft or 4 ft 6 in as the nearest standard size.

The length of a transfer unit is given by

$$\begin{aligned} L_t &= 0{\cdot}1 c_p G^{0{\cdot}84} D \\ &= 0{\cdot}1 \times 0{\cdot}24 \times 260 \times 4{\cdot}5 \\ &= 26 \text{ ft} \end{aligned}$$

and as there will be 1·75 transfer units the length of the dryer will be 45 ft 6 in, which gives a ratio of length to diameter of approximately ten, which is acceptable.

The ingoing air has a humidity of 0·008 lb/lb so that the ingoing humidity will amount to $11{,}850 \times 0{\cdot}008 = 94{\cdot}8$ lb/h. As the total moisture removed from the feedstock is 333 lb/h, the exit air will pass out 427·8 lb/h, which gives the exit gases a humidity of 0·0361 lb/lb at 136°F or a relative humidity of about 31% and a specific volume of 17 ft^3/lb. The exhaust fan will therefore have to exhaust (11,850 × 17)/60 = 3,350 ft^3/min.

The dryer would be run at about 30/4·5 = 6–7 rev/min, would have a slope of $\frac{1}{4}$ inch in ten and would have 14 flights which would be about 6 in in height. They would be provided with a fairly liberal 125° lip. The volume of the dryer will be 725 ft^3 and, assuming an 8% hold-up with material of 50 lb/ft^3 bulk density, the hold-up will be 2,900 lb and the residence time will be 2,900/1,000 = approximately 3 hours.

The steam heater battery will have to raise 11,850 lb/h of air from 60°F to 290°F calling for

$$0{\cdot}24 \times 11{,}850 \times (290-60) = 655{,}000 \text{ Btu/h}$$

and, as the latent heat of condensation of 150 lb/in^2 gauge steam is 857 Btu/lb, the steam requirement will be 758 lb/h, giving a specific steam consumption figure on 333 lb moisture evaporated per hour of 2·28 lb/lb of water evaporated. The evaporation/volume ratio on the dryer would be 333/725 = 0·46 lb of water evaporated per ft^3 of dryer volume. The sensible heat losses in the exhaust air would be about 216,000 Btu/h representing 33 per cent of the heat input at the steam heater.

Indirect Rotary Dryers

There are certain instances where direct-heated rotary dryers are not suitable for a particular operation. These will generally fall within the following categories:

(1) Where the material undergoing drying cannot be exposed to combustion gases.
(2) Where excessive dust carry-over will occur through the entraining effect of air or gases in a direct-heated dryer.
(3) Where low-cost steam is available, usually at low to medium pressures.
(4) Where volatile components are present which justify their recovery.

In such cases it is usual to employ one or other of a variety of indirectly-heated rotary dryers. In indirect dryers, the heat required for heating up the feedstock and evaporating its moisture content is supplied essentially by conduction through the dryer shell or from internal heating tubes, usually supplied with steam. In the former, the dryer shell is usually housed in a refractory-lined chamber or external concentric shell and the combustion gases, after being led around the shell, are either exhausted directly to atmosphere or may be led through an internal cylindrical shell and thence exhausted. Fig. 37 shows such a dryer in its simplest and least efficient form, whereas Fig. 38 shows another arrangement, consisting of an external concentric shell furnished with combustion gases from a combustion

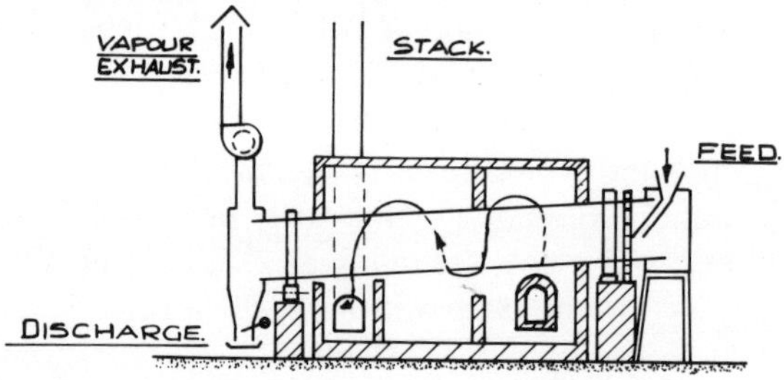

Fig. 37. Indirect-heated rotary dryer.

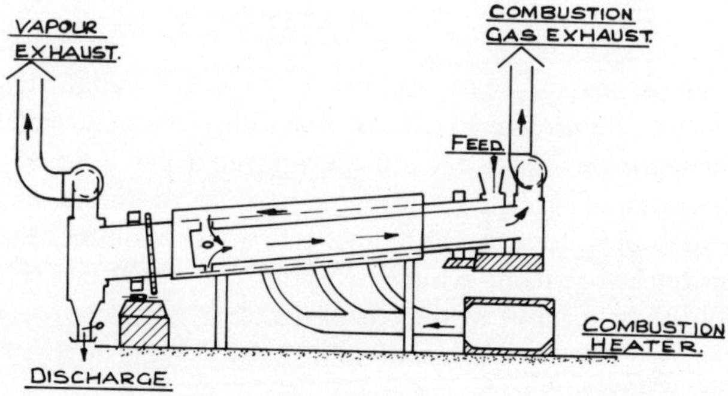

Fig. 38. Double-shell indirect-heated rotary dryer.

chamber, these gases being then led through an internal tube before discharging to atmosphere.

A more efficient design is shown in Fig. 39 and Plate XXIII. In this, products of combustion are led into an internal tube from which they pass by a connecting duct to hollow lifting flights and thence to atmosphere. In this design it will be seen that the raw feedstock passes through the dryer co-current with the combustion gases but not in contact with them, whilst the combustion gases after leaving the inner tube pass counter-current to the movement of the feedstock and out of contact with it,

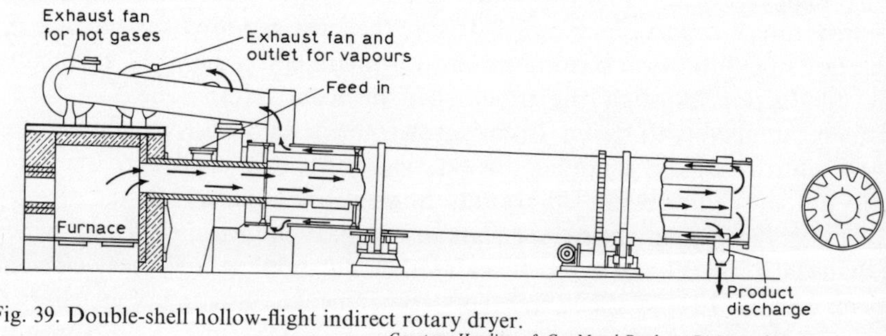

Fig. 39. Double-shell hollow-flight indirect rotary dryer.

Courtesy Hardinge & Co. Metal Products Division of Koppers Co. Inc.

Plate XXIII. Double-shell indirect rotary dryer.
Courtesy Hardinge & Co. Metal Products Division of Koppers Co. Inc.

through hollow lifting flights. By this means, the maximum possible area of contact is made available, feedstock is brought into contact with the hottest part of the inner tube, whilst the exhaust gases run counter-current with the feedstock, thereby undergoing a major reduction in temperature. Such a dryer results in exceptionally high thermal efficiencies.

Heat transfer in these dryers is entirely by conduction and radiation—the latter attaining important proportions depending on the surface temperature. Direct contact with air or a neutral gas is restricted to that amount required to remove liberated moisture, and heat transfer from this source is usually of a negative nature. Unless the feedstock is excessively light and dusty, the dryer shell and, where used, the inner tube will be provided with lifting flights both to increase the heat transfer area and to assist drying by the showering effect and disturbance of the feedstock in contact with the heated surfaces. Where the material undergoing drying may only be brought into contact with surfaces at steam temperature the most usual form of dryer is that shown in Fig. 40, known as a steam-tube dryer.

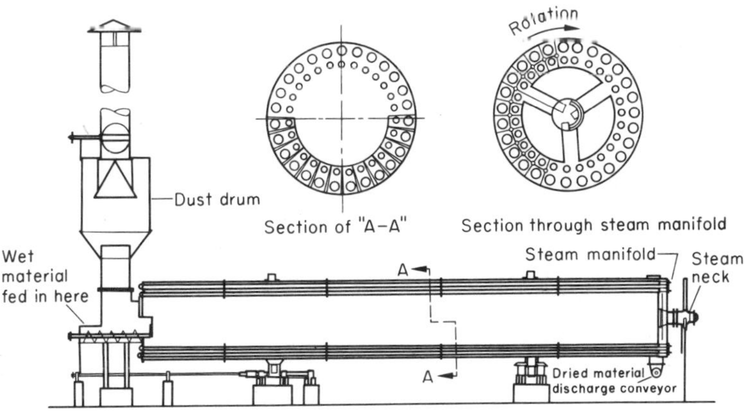

Fig. 40. Indirect steam-tube (Louisville) rotary dryer. *Courtesy Constantin (Engineers) Ltd.*

In this dryer, steam-heated tubes in one or several peripheral banks running the whole of the length of the dryer are secured to the dryer shell, rotate with the latter, ploughing through and cascading the feedstock during its passage through the dryer. The number of banks will depend largely on the physical nature of the feed, the more sticky the latter, the fewer the heater banks. Lifting flights may also be fitted to improve the cascading effect. Movement of the material through the dryer is promoted by the slope of the dryer from the horizontal, and hold-up in the dryer and, hence, retention time are usually taken care of by means of radial discharge chutes or retainer dam plates within the dryer at the discharge end. Steam supply to and condensate removal from the steam heater tubes is through a specially designed trunnion end and connection branches. Continuous venting of non-condensable gases is usually provided for at the other end of the tubes. A gas exhaust stack or fan removes moisture vapour as it is liberated from the feedstock. The thermal efficiency of this type of dryer is exceptionally high, of the order of 75–90%, specific steam consumptions ranging from 1·5–1·8 lb steam/lb water evaporated.

A more simple variant of the steam-tube dryer suitable for free-flowing materials is shown in Fig. 41.

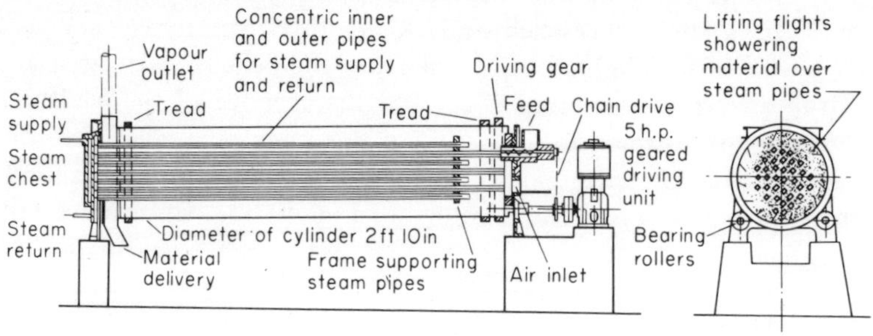

Fig. 41. Fixed-tube rotary dryer. *Courtesy A.P.V.-Mitchell (Dryers) Ltd.*

In this, the shell, fitted with lifting flights, rotates whilst a steam-heated basket of tubes, usually of square cross-section, remains stationary. Material showered by the lifting flights on the inner surface of the shell cascades over the heater tubes and, due to the slope of the dryer, moves from the feed end to the discharge end.

The use of indirect rotary dryers operating at atmospheric pressure is, of course, restricted to materials which can withstand the surface temperatures produced. Where these have to be low recourse may have to be made to vacuum drying if the dustiness, susceptibility to the presence of combustion gases or air, or the presence of volatile or toxic components dictate the use of an indirect-heated dryer. Indirect dryers, other than steam-tube dryers, may have to be operated in a similar manner to a kiln when the material is excessively dusty. In such cases, lifting flights may have to be omitted altogether or may be of very meagre dimensions to promote a turning-over action to the charge.

Transfer of heat, being by conduction, and possibly by radiation, will vary widely depending on the temperature of the heating surfaces. The latter will vary from that of steam in the case of steam-tube dryers to as high as 1,500–1,800°F for fuel-fired dryers. For high-temperature operation the dryer shell will usually be constructed in heat-resisting alloy steel. Heat-transfer rates ranging from 2–10 Btu/h/ft^2/deg F for steam-heated dryers to 100–400 Btu/h/ft^2/deg F for fuel-fired indirect rotaries are met with where combustion gas temperatures are high and the effect of radiant heat is high. Material hold-up in indirect rotary dryers is usually arranged to be somewhat higher than in direct rotaries, being of the order of 5–15% of the dryer volume as compared with 2–10%, but the ratio of length to diameter is usually common to both.

The drying cycle characteristics of materials dried in an indirect rotary dryer will be essentially the same as in direct dryers but the temperature of the material will more nearly approach the boiling point of the moisture during the constant rate period. The drying rate will be governed by the rate of heat transfer from the hot combustion gases in fuel-fired dryers or condensing steam in steam-tube dryers in accordance with the relationship $q = UA_e(\Delta t)_m$ where q = heat transferred in Btu/h, U = overall heat transfer coefficient, Btu/h/ft^2/deg F, A_e = effective drying area, ft^2, and $(\Delta t)_m$ = the mean temperature difference between the material and the heating medium. However, in an indirect rotary dryer, heated by combustion gases, whether lifting flights are provided or not, the value of A_e is difficult to assess and it would be necessary to determine a value for UA_e from pilot-scale tests on the type and design of dryer to be employed, unless it were available from past experience on similar feedstocks. Most of the resistance to heat transfer will be in the material itself—depending upon its moisture content—and between the material and the heated surfaces. The latter are likely to be at a temperature close to that of the heating medium and the heat-transfer rate will be materially governed by the degree of agitation of the material in its contact with heated surfaces. Evaporation rates in fuel-fired indirect rotaries will usually range from 1 lb/ft^3/dryer volume/h in single-shell dryers to 2·8 lb/ft^3/h for double-shell dryers fitted with hollow lifting flights. Steam-tube dryers will give evaporation rates ranging from 0·5–2·5 lb/h/ft^2 of steam-tube area.

The mechanical features associated with indirectly-fired rotary dryers are similar to those employed with direct rotary dryers. Girth rings and gear rings will, however, have to be mounted substantially at the ends of the dryer away from the heated length of the dryer. Exhaust fans for single-shell dryers will only have to remove moisture vapour and enough air to keep the dryer purged; exit air will therefore be at a high humidity so that adequate insulation of the exhaust duct and fan is necessary to prevent re-condensation. In double-shell indirect dryers, whether with plain or hollow lifting flights, the exhaust fan will have to be of adequate capacity to handle the combustion gases leaving the dryer together with the moisture vapour. Should there be any dust entrainment in single-shell dryers it may be necessary to install a wet dust scrubber; humidities will probably be too high to permit of the use of cyclones or bag collectors. Drying conditions will usually be controlled from the discharge temperature of the product. Very wet or pasty feedstocks may tend to build up on surfaces at the feed end and it is usual in such cases to back-mix dry product with the feed. As the degree of back-mixing may be considerable, particularly with material which is at all sensitive to high temperatures calling for somewhat lower hot gas inlet temperatures, not only has all the requisite conveyor equipment and back-mixer to be installed, but the physical dimensions and, hence, cost of the dryer may have to be large in order to provide the necessary residence time for drying. Fuel consumption will also be higher as a result of having to re-heat the dry material to its original discharge temperature. Wherever possible, therefore, it is advisable to consider very carefully whether the material can be dried in a parallel-flow direct rotary without back-mixing as a more economical solution.

Indirect-direct Rotary Dryers

Extremely high thermal efficiencies are obtainable in rotary dryers which embrace both indirect and direct heating. As, however, the feedstock necessarily comes into contact with high-temperature surfaces they are not usually suitable for heat-sensitive materials. Moreover, there is some danger of dust entrainment and they are therefore not very suitable for light, dusty materials.

The simplest form of indirect-direct rotary dryer is shown in Fig. 42.

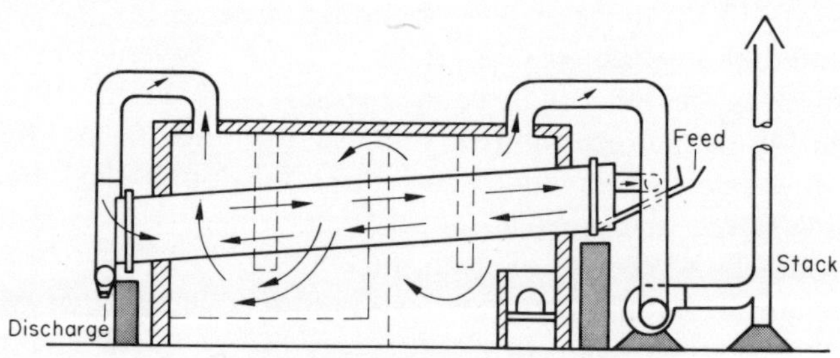

Fig. 42. Single-shell indirect-direct rotary dryer.

In this design, combustion gases from an external or built-in combustion chamber are led around the dryer in a refractory brick housing or a concentric outer shell refractory-lined internally, thus transmitting heat through the dryer shell. Spent hot gases are then led directly into the dryer shell either parallel or counter to the flow of feed and thence exhausted to atmosphere. Either part or the whole of these gases can pass through the dryer as desired by adjustment of dampers. In Fig. 42 the gases are shown moving counter-current to the feed. Heat transfer from the enclosure is by radiation, convection and conduction through the dryer shell to the material, followed by direct convection heat transfer from the gases to the material as they pass through the dryer. For high moisture feeds or for very low exit moisture contents, it will be counter to the flow of material.

The most usual design employed today is that shown in Fig. 43, which is a double-shell indirect-direct rotary dryer.

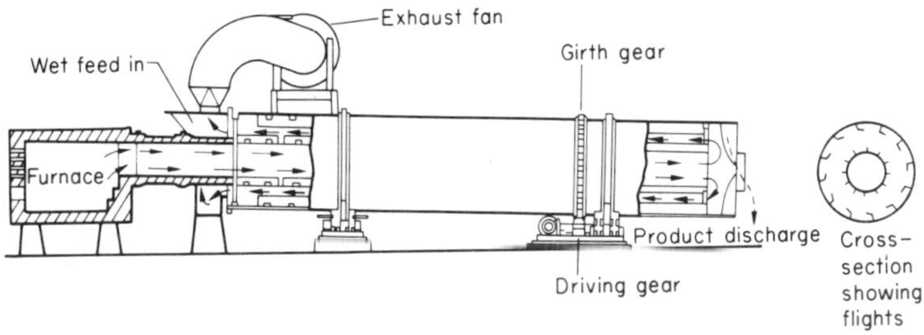

Fig. 43. Double-shell indirect-direct rotary dryer.
Courtesy Hardinge & Co. Metal Products Division of Koppers Co. Inc.

In this design combustion gases from a combustion chamber are led to an inner tube running practically the whole length of the dryer and provided with external lifting flights. Combustion gases leaving the inner tube are led through an annular space between this and the main outer shell of the dryer which is provided with internal lifting flights in the usual manner. Spent combustion gases are exhausted by an adequate stack or preferably by an exhaust fan. The dryer therefore consists essentially of an indirect parallel-flow dryer combined with a direct counter-flow dryer. Dried product is discharged through the action of special lifting flights combined with a discharge chute. Indirect-direct rotaries usually have high thermal efficiencies but, where corrosion or abrasion occurs, are difficult to maintain due to internal restrictions of space. Due to the considerable surface area of the inner shell, usually at an elevated temperature, and the intimate counter-flow of material and gases in the annular space, heat-transfer rates may be 30–35% higher than in a single-shell direct dryer of the same volume. Evaporational capacities 2·5–9·0 lb/ft^3 total dryer volume are usual with thermal efficiencies of 60–80%.

The thermal efficiency of a dryer is the ratio of the theoretical heat required to remove the moisture during drying to the total heat delivered to the dryer expressed as a percentage. Whereas the theoretical requirement is that heat is only consumed in evaporating the moisture, the total heat consumption will be this together with the heat to raise the solids content of the feedstock from its initial temperature to the temperature at discharge; the heat in the waste humid exhaust air leaving the dryer as sensible heat (i.e. excluding the latent heat of the vapour content) above the fresh air inlet temperature; and the heat losses resulting from radiation, convection and conduction to the surrounding air, foundations, etc. The heat required to raise the feedstock solids from feed temperature to discharge temperature is a necessary consumption in the operation and cannot be avoided. If the discharge temperature is very high, some heat may be recovered by preheating the air requirements in a solids-air continuous cooler. Heat losses in the exhaust air and as a result of radiation, convection and conduction can be very high and may be to some extent avoidable, with consequent savings in fuel costs. The former can be controlled by proper operation of the dryer, whilst the latter can be materially reduced through the use of adequate thermal insulation.

Exhaust Gas Heat Losses

The major heat losses in a rotary dryer installation are usually those in the exhaust gases. Such losses in an indirect dryer will be largely controlled by the design of the dryer, assuming excessive temperatures are not employed, and are not very amenable to reduction. The use of the exhaust gases as a purging vehicle within the dryer to whatever extent is permissible will assist in improving the efficiency of the dryer. In direct and indirect-direct dryers, however, these losses can be reduced by proper operation of the dryer. Any cold air leakage into the dryer at feedstock inlet and at end-hood seals results in high exhaust losses as does excessive secondary, tertiary or diluent air fed to the combustion chamber. The heat required to raise air initially at 60°F and 40% relative humidity to different temperatures is shown in Fig. 44 in terms of Btu/lb, and as a convenience the volume which this air assumes at these temperatures as ft^3/lb is also given.

Air leaving a direct rotary dryer will of course have picked up moisture in drying the feedstock and its absolute humidity will have increased accordingly. Whereas the sensible heat losses increase with increases in temperature at constant absolute humidity, this does not strictly apply where the air picks up moisture and increases its humidity. The degree of saturation has an important bearing. The sensible heat loss, although directly proportional to temperature increase, is inversely proportional to

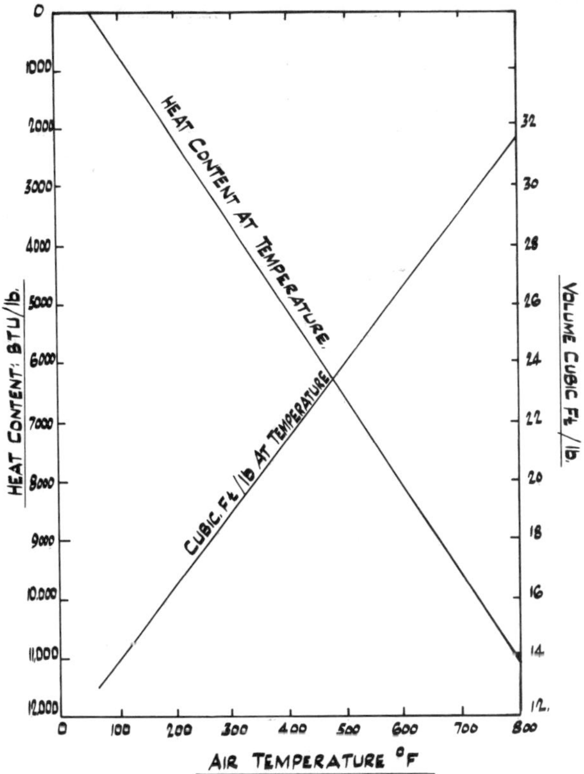

Fig. 44. Heat content of air at different temperature.

increase in humidity. With increase in humidity the heat loss decreases with increase in temperature. At a given temperature the sensible heat loss varies approximately inversely as the percentage saturation increases, approaching a minimum as the humidity approaches saturation. This is illustrated in Fig. 45 where sensible heat losses above 70°F in Btu/lb of water evaporated are plotted against saturated air exit temperatures. As evaporation of moisture from material dried in a direct rotary dryer is carried out substantially at atmospheric pressure, a value of 1,150 Btu/lb can be assumed for the heat requirement, and it will be appreciated from Fig. 45 that, as the exhaust air increases in temperature and humidity, the sensible heat losses fall from a maximum of around 20 per cent to a minimum of around 5 per cent for saturated air at an exit temperature of 190°F.

For practical purposes, the exhaust gases from a direct rotary dryer should be withdrawn at as high a humidity as possible but at temperature well above dew-point so as to prevent condensation in subsequent ducts, cyclones and fan. In indirect rotary dryers, the re-passage of combustion

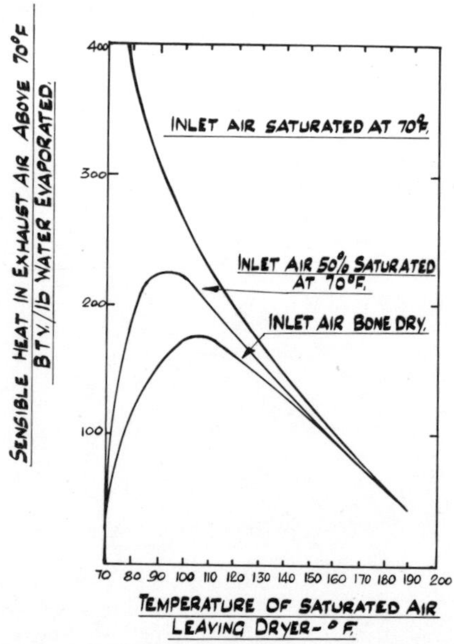

Fig. 45. Heat losses in exhaust air v. water evaporated.

(after Perry, I. H. Chemical Engineers Handbook, 2nd Edn. 1941)

gases through internal tubes or hollow lifting flights will considerably increase the thermal efficiencies of the dryer by recovering heat and thus reducing the exit temperature of the gases.

Radiation, Convection and Conduction Heat Losses

All rotary dryers, other than indirect dryers the shell of which is housed in a heated chamber itself adequately insulated, will suffer considerable heat losses through convection and radiation, the extent of which will depend upon the shell surface temperature and that of the surrounding air. In direct rotaries, employing steam battery heated air, and indirect jacketed rotaries, surface temperatures will not be very high but will still justify external thermal insulation. Fuel-fired direct rotary dryers will suffer very considerable losses both from the combustion chamber surfaces and from the dryer shell surface unless adequate attention is paid to insulation. In the case of the dryer shell, this could mean the difference between a heat loss of 5,000–6,000 Btu/ft^2/h when uninsulated and 250–300 Btu/ft^2/h if adquately insulated. The direct-fired air heater should not only have internal refractory and insulating brick but should have the secondary, tertiary or diluent air fed to the combustion chamber through an annular space surrounding it, thus improving combustion and utilising as much heat passing through the chamber insulation as possible.

Heat losses from the shell are reduced by lagging the shell surface, there being several methods and materials for doing so depending upon the severity of heat losses to be anticipated and the cost of lagging considered justifiable. Taking as an example a normal surface temperature of a dryer shell 6 ft diameter and 35 ft long, the cost of heat lost if uninsulated with the dryer fuel oil fired could amount to as much as £400 to £500 per annum, which would represent a considerable and unnecessary increase in the operating cost of the dryer. The heat loss by natural convection from a hot surface to ambient air can be approximately calculated from the expression

$$h_c = 0{\cdot}7 + \frac{\Delta t}{375}$$

where h_c = heat transfer coefficient in Btu/ft²/h and Δt is the temperature difference between the surface and that of the ambient air in °F. Heat lost by radiation is given approximately by

$$Q = bAeT^4$$

where Q = heat lost in Btu/h; A = surface area in ft²; T = the absolute temperature of the surface in °F; e = a correction factor for "blackness" of surface, usually assumed to be 0·8 for mild steel; and b = a constant which may be taken as having a value of $17{\cdot}3 \times 10^{-10}$. For practical purposes the expression becomes

$$Q = 17{\cdot}3 \times 10^{-10} \times 0{\cdot}8T^4 \text{ in Btu/ft}^2\text{/h}$$

Heat lost by a combination of convection and radiation is therefore

$$\left(0{\cdot}7 + \frac{\Delta t}{375}\right) + 17{\cdot}3 \times 10^{-10} \times 0{\cdot}8$$

in Btu/ft²/h and for practical purposes the expression $C(t_1 - t_2)^{0{\cdot}25}$ may be substituted for

$$\left(0{\cdot}7 + \frac{\Delta t}{375}\right),$$

where t_1 is the temperature of the hot surface (°F), t_2 that of the ambient air and C is a constant which for large cylindrical shapes such as a rotary dryer shell can be taken as approximately 0·3.

In practice, of course, the surrounding air will not usually be still and convection losses will be higher than those assumed from the above equation. Strong draughts or convection currents may result in 50–100% increase in overall losses. The insulation of the dryer shell is usually carried out either by applying glass or mineral wool blankets strapped around the shell; by applying asbestos or similar pulp to wire-netting wrapped around the shell; or by strapping on magnesia or other slabs and covering their surfaces with a trowelled asbestos or other preparation, finally painting the whole of the exterior with white paint. Table 20 gives the thermal conductivity of materials usually employed over a range of surface temperatures.

Table 20. Thermal conductivities of insulating materials, Btu/lb/ft^2/deg F/ft

Material	*Mean temperature, °F*							
	100	200	300	400	500	600	800	1,000
Asbestos felt laminations	0·033	0·037	0·04	0·044	0·048			
Asbestos	0·097	0·110	0·117	0·121	0·123	0·125	0·130	
85% magnesia	0·039	0·041	0·043	0·046				
Cork	0·026	0·052						
Rock wool	0·030	0·034	0·039	0·044	0·05	0·057		

Performance of Rotary Dryers

Typical evaporative capacities of rotary dryers over the ranges usually experienced in practice are given in Table 21.

Table 21. Evaporative capacities of rotary dryers

Type of dryer	*Evaporative capacity range lb/ft^3 dryer volume*
Direct: steam-heated	0·15–2·0
fuel-fired	1·8 –7·5
Indirect-direct: fuel-fired	2·5 –9·0
Indirect: fuel-fired (single-shell)	1·0 –1·8
fuel-fired (double-shell)	1·4 –2·5
Steam tube	3·0 –7·5
Steam-jacketed vacuum	0·5 –1·5 lb/h/ft^2 heated surface

The performance of typical commercial installations of the different types of rotary dryer for a variety of products is given in Tables 22 and 23.

Table 22. Performance of direct-heated rotary dryers

	Sugar beet pulp[1]	*Calcium carbonate*[1]	*Blast furnace slag*[1]	*Lead concentrate*[2]	*Sand*[2]	*Zinc concentrate*[2]	*Ammonium sulphate*[3]	*Fine salt*[3]	*Crystals*[4]	*Chemicals*[4]
Air flow	Parallel	Parallel	Parallel	Parallel	Parallel	Parallel	Counter	Counter	Counter	Indirect counter
Dryer diameter	9 ft 2 in	6 ft 3 in	7 ft 2 in	4 ft 5 in	4 ft 6 in	7 ft 6 in	9 ft 0 in	5 ft 0 in	10 ft 0 in	4 ft 6 in
Dryer length	46 ft 0 in	34 ft 0 in	40 ft 0 in	35 ft 0 in	32 ft 6 in	60 ft 0 in	40 ft 0 in	40 ft 0 in	60 ft 0 in	27 ft 0 in
Method of heating	Oil	Oil	Oil	Oil	Gas	Oil	Gas	Steam	Steam	Louisville steam tube
Method of feed	Screw	Belt	Belt	Screw	Chute	Screw	Conveyor	Feeder	Screw	Screw
Initial moisture, % w/w	82	13·5	33	14	5·65	18	2·5	5·0	7·0	1·5
Final moisture, % w/w	10	0·5	Nil	8	0·043	8	0·2	0·1	8·99	0·1
Evaporation, lb/h	34,000	6,000	11,600	1,393	701	8,060	1,120	400	1,150	63
Capacity, lb evaporated/ft³ dryer volume	11	6	7	2·5	1·35	2·3	0·5	0·52	0·245	—
Efficiency, Btu supplied/water evaporated	1,420	1,940	1,710	2,100	2,550	1,850	1,920	2,100	1,650	—
Inlet air temperature, °F	1,560	1,560	1,560	1,300	1,650	1,500	400	280	302	—
Outlet air temperature, °F	230	220	248	200	222	200	180	170	144	—
Residence time, av. min	20	25	30	20	12	20	15	40	70	30/40
Fan hp	70	40	50	20	5	75	25	8	—	10
Motive hp	15	20	25	10	10	55	60	15	60	—
Fan capacity, std air ft³/min	45,000	8,500	18,000	2,750	2,100	12,000	18,500	6,500	—	—

[1] *Courtesy Buell Ltd.*
[2] *Courtesy Head Wrightson (Stockton) Ltd.*
[3] *Courtesy Edgar Allen Aerex Ltd.*
[4] *Courtesy Constantin Engineers Ltd.— Louisville Dryers.*

Table 23. Performance of indirect-direct rotary dryers

	Indirect-direct double-shell		
	Coal	*Anhydrite*	*Coke*
Dryer diameter	7 ft 6 in	5 ft 10 in	5 ft 10 in
Dryer length	46 ft 0 in	35 ft 0 in	35 ft 0 in
Initial moisture content, % w/w basis	22	6·0	15
Final moisture, % w/w basis	6	1·0	1·0
Evaporation rate, lb/h	5,800	2,300	1,600
Evaporation – volume ratio lb/ft^3/h	3·5	3·15	2·2
Heat source	Coal	Oil	Oil
Efficiency, Btu supplied/lb water evaporated	1,250	1,250	1,340
Inlet air temperature, °F	1,200	1,350	1,350
Outlet air temperature, °F	160	160	200

Courtesy Edgar Allen Aerex Ltd.

CHAPTER VIII

Flash or Pneumatic Dryers

If a material is reasonably free-flowing and of not too large a particle size, it may, by introduction to a flow of air at adequate velocity, become totally dispersed and entrained in such a manner that it will be pneumatically conveyed from the feed point to some other point of delivery. If this air stream is hot and the material wet, drying will take place and the dried product can be separated from the air by a suitable reduction in velocity. A flash or pneumatic dryer is essentially a dryer which is designed to secure this effect in the most efficient manner possible. At the same time it may be designed to act as a pneumatic conveyor to convey the feedstock from its point of production in the moist condition to a convenient point where it can be stored or packed in its dried condition. Where the feedstock is initially in a suitable condition for entrainment, no preliminary treatment is required, otherwise it may require preliminary disintegration or preforming or it may need the addition of some dried product in a back-mixer prior to feeding into the system. In certain designs the hot air or gases are introduced into a disintegrator in which either partial or complete drying occurs. Separation of dried product from the conveying air usually takes place in a cyclone followed by a further final separation in cyclones, fabric sleeve filters or wet scrubbers as may be necessary. A typical pneumatic dryer is shown in Fig. 46 and an industrial installation is illustrated in Plate XXIV.

A pneumatic drying plant consists essentially of a source of hot gases, either hot air or combustion gases produced by a steam or electric heater, an indirect fuel-fired heat exchanger or a fuel-fired direct-heated air heater combustion chamber; a material feeding device; the main drying column or duct usually provided with a venturi section at the material feed point; a cyclone for material-air separation in which the final stage of drying may take place and fitted with a rotary discharge valve; and an air exhaust

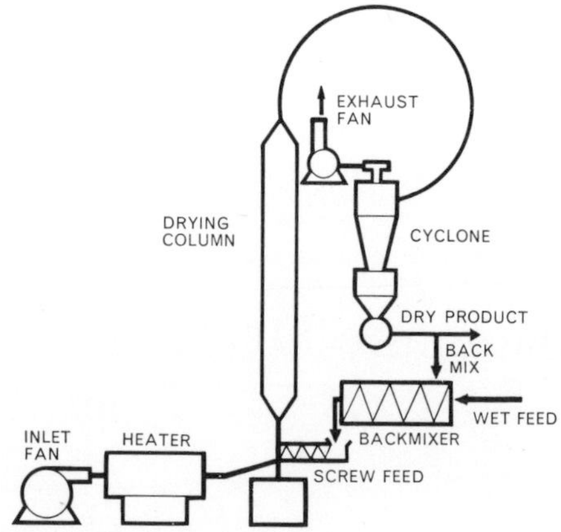

Fig. 46. Diagrammatic sketch of pneumatic dryer.

Plate XXIV. Berk ring dryer installation. *Courtesy Pennsalt Ltd.*

fan for discharging the humid air either to atmosphere or to any subsequent dust recovery equipment found necessary. The source of hot gases needs no special mention as it will take the form of those used on direct-heated rotary dryers. However, in view of the quantity of hot gases required, it may be necessary in drying high purity products to employ a suitable filter to clean the air. The feeding arrangements will also take the form of those suitable for rotary dryers, wherever possible interposing a rotary valve between the feeder and feed point to prevent leakage of air into or out of the dryer. Where back-mixing of dried product is necessary this will be located before the feeding device. When a true venturi is fitted to the drying column it may be necessary to have a positive pressure on the up-stream side, in which case material may be fed into the eye of a paddle-blade design of fan of robust construction which may act as both a fan and mild disintegrator. Some designs have the venturi in a horizontal section of the main duct system whereas others have it in a vertical section. The use of a venturi assists in both completing entrainment locally and in producing a point of suction to assist feeding where no positive mechanical feeding device is called for.

The main conveying and drying duct may be of circular or rectangular cross-section but in either case should be extremely well finished internally and free from projecting edges which can interfere with air flow or which may cause build-up of material. The main cyclone should be similarly made with close attention to smooth surfaces and joints. Inspection and cleaning ports in the duct are an added advantage. The degree of drying which takes place in the main ducting as opposed to within the cyclone depends upon the individual manufacturer's design of dryer. In some designs or under certain circumstances either, or both, air and material recirculation may be employed with a view to improving the thermal efficiency in the case of the former, or the uniformity of drying in the case of the latter. Properly designed throughout, based on carefully executed pilot-scale tests, dryers of this type can accomplish a drying operation in an extremely efficient manner with very close control of the drying conditions to which the feedstock is continuously subjected, and at less capital cost than many other types of dryer. The time of contact between material and hot gases is a matter of seconds only so that heat-sensitive materials having good drying characteristics are particularly suitable for handling in this type of dryer. On the other hand, materials which should not be subjected to conditions liable to cause attrition or which are of a sticky or greasy nature may lead to detrimental effects to the dry product or to operational difficulties. Due to the almost instantaneous heat transfer from hot gases to moist feedstock, high inlet gas temperatures can be applied to heat-sensitive materials which will be subjected to drying temperatures for an exceedingly short period of time during which they attain a temperature never greater than the gas outlet temperature, these being completely parallel-flow dryers. Where the material is both heat-sensitive and has a relatively long drying cycle it may be necessary to employ more than one drying system in series at appropriate inlet gas temperatures.

Pneumatic Dryer Design

A description of the different types of pneumatic dryer is given later in this chapter and the designing and sizing of such dryers will largely depend on experience and the results of pilot-scale tests on a unit as closely simulating the proposed installation as possible. Assuming adequate thermal insulation – not difficult to achieve in this type of dryer – the mechanism of drying will closely approach that of a parallel-flow adiabatic dryer. Pilot-scale tests will quickly establish the air velocity required to entrain the feedstock completely and maintain it in a state of pneumatic suspension. They will also establish the maximum inlet dry-bulb gas temperature permissible without detriment to the product. Where the feedstock has mostly free moisture, its temperature will never be appreciably above the wet-bulb temperature of the air with which it is in contact. Basically the operation is a combination of heat-transfer considerations, psychrometry and pneumatic handling technology. From the point of view of pneumatic handling the ratio of solids to gas weights will usually be between 0·1 and 1 lb of solid per lb of gas, a figure of not less than 0·5 being a satisfactory average, and gas velocities within the duct system will lie between 40 and 150 ft/s with a value of 40–80 ft/s for most materials dealt with in this type of dryer measured at N.T.P.

For a total-rejection straight-duct flash dryer, making certain assumptions, the drying time and hence the length of duct, ignoring any drying effect in the cyclone, can be calculated from first principles, and hence, if the gas velocity is known, the duct length. It is, however, necessary to assume that practically all the moisture is surface moisture, that the material particles are spheres and that a value can be given to the coefficient of heat transfer in the gas film. The heat equation for the process will be as usual:

$$q = U_s A (\Delta t)_m \text{ in Btu/h}$$

where $(\Delta t)_m$ is the logarithmic mean temperature difference between the material and gas at feed point and discharge from the duct, A is the particle surface area in ft^2 and U_s is the surface heat transfer rate. The total heat requirement, assuming no heat losses and ignoring sensible heat gained by the material, will be:

$$Q = d_s V \lambda W$$

where d_s is the particle density in lb/ft^3, V is the particle volume in ft^3 and W is the weight of moisture removed per lb bone-dry solid. As $V = \pi D^3/6$ for a sphere of diameter D ft, the drying time will therefore be in hours:

$$\theta = \frac{Q}{q} = d_s \frac{V\lambda W}{U_s A(\Delta t)_m} = d_s \pi \frac{D^3 \lambda W}{6U_s A(\Delta t)_m}$$

The value of U_s is of course unknown, but under conditions normally experienced in pneumatic dryers, may be taken to be approximately equal to $2k_f/D$, where k_f is the thermal conductivity of the gas film in Btu/h/ft^2/°F/ft, whence

$$\theta = \frac{d_s D^2 \lambda W}{12k_f(\Delta t)_m} \text{ in hours}^{(6)}$$

and the length of the drying duct in feet will be θv, where v is the velocity of the drying medium in ft/h.

The above method of estimating the drying time and hence the length of the drying duct is highly theoretical and would be unreliable for design purposes unless supported by pilot-scale tests. In practice, the material subjected to drying is rarely composed of spherical particles and may have a fairly wide screen analysis. The moisture content is frequently not restricted to surface moisture and, due to the range of size of particle, the residence time will vary widely from particle to particle. The actual velocity of particles will range between rather more than the terminal velocity of larger and heavier particles and that of the lightest particles which may be travelling at the velocity of the air stream. The velocity of the air stream must be above the terminal velocity of fall of the heaviest particles so that the general movement of all the feed is upwards through the duct with no fall-back of material. This must be determined by tests in the pilot-scale plant. In practice, duct bends and the cyclone itself constitute an additional factor of safety above the calculated residence time and should be of the order of upwards of 200 per cent. For this reason it is not unusual to employ a high efficiency cyclone design of at least twice the volume of the proposed duct. Additional classification of the material takes place within the cyclone, the largest particles following the circumferential sides of the cyclone while the lightest particles take a shorter path. The resultant differential drying times result in more uniform drying of the material. Where there is considerable disparity of particle size it is quite usual to use two cyclones in series to act as classifiers and to recycle a proportion of the heavier particles with the feedstock either in a back-mixer or at a venturi feed point where the considerable velocities existing within the venturi throat cause intimate mixing and consequent rapid equalisation of moisture content in the feed at this stage.

Different dryer manufacturers have different methods of arriving at the size and operating method of commercial scale flash dryers depending upon the system employed in the particular design and their scale-up expertise. In all cases, however, it is imperative to carry out pilot-scale tests from which the following data will be obtained at optimum yields and correct final moisture contents:

Air velocity to convey material pneumatically, ft/s
Feed rate, lb/h
Product rate, lb/h
Initial moisture content, % by weight
Final moisture content, % by weight
Ambient air temperature, °F
Ambient air humidity, lb/lb
Inlet air temperature, °F
Outlet air temperature, °F
Material feed temperature, °F
Material discharge temperature, °F
Percentage solids recovery
Material recycle ratio
Air recycle ratio

Familiarity with the test unit will indicate heat losses by convection and radiation under optimum operating conditions together with anticipated heat losses on full-scale plant under the same conditions. The total heat requirements of the dryer will be the sum of the heat taken up by the solids, the heat to heat up the moisture content to the exit air temperature and to evaporate the moisture, and the heat losses from the dryer. This heat is supplied by the cooling of the hot air feed from its entry to its exit temperatures, whence the quantity of air required. Experience will dictate the scale-up factor required for air quantities as well as length of ducting and size of cyclone. With very low moisture content feeds the quantity of air arrived at from the above considerations may be lower than that required to give satisfactory entrainment and pneumatic conveyance, in which case an allowance must be made. Good practice indicates that the ratio of weight of air to weight of material should not be less than 2 : 1. Where venturis are used these are usually designed to operate at 60–80 ft/s internal air velocities and for good practice the inlet velocity to the cyclone should be a minimum of 70 ft/s.

A somewhat different and more scientific method of approach in designing a pneumatic dryer combined with scale-up experience is sometimes employed. When a drying operation such as that encountered in a pneumatic dryer is carried out in a truly adiabatic manner, the latent heat of evaporation is exactly provided by the temperature drop of the air stream and the total enthalpy of the air stream including moisture vapour released during drying remains constant. Under these conditions, which are ideal and rarely occur in practice, due to heat losses, the wet-bulb temperature of the air stream remains constant and the condition of the air stream follows a simple wet-bulb line on a psychrometric chart. Under these ideal conditions the amount of moisture picked up between inlet and outlet air conditions is given on the psychrometric chart and automatically takes into account latent heat requirements. In practice, however, heat losses due to convection and radiation occur whilst, at the same time, heat has to be supplied to raise the solids and moisture from inlet to outlet temperature. From the evaporational standpoint these can be regarded as heat losses on the system and have to be supplied by the air stream. Instead of moving along an ideal wet-bulb line on the psychrometric chart, the air conditions will follow a line at a flatter angle to the horizontal and for the same exit air temperature will terminate at a lower wet-bulb temperature and will contain less moisture than under ideal conditions. It is found that the angle between the ideal wet-bulb line and the actual operational cooling line is a trigonometric function of the heat losses divided by the weight of water evaporated and is referred to as the heat loss factor. The value of this angle can therefore be calculated for different known heat loss factors. There is also a trigonometric relationship between this angle and the angle of inclination of the ideal wet-bulb line to the vertical. As the value of the wet-bulb offset angle traverses its range corresponding with different values of the ideal cooling curve inclination, the value of the heat loss factor ranges from minus infinity to plus infinity. If an arbitrary point be taken on an enthalpy-humidity chart (see Fig. 5) a convenient scale parallel with the enthalpy scale can be constructed for a range of values for the heat loss factor from zero upwards using corresponding values for the offset angle.

Pilot-scale tests will establish the sensible heat in the products and ambient and inlet hot air temperature and humidity conditions. They will also establish the exit air dry-bulb temperature for a specified weight of water evaporated. For the desired production plant the heat loss factor will be the sum of heat passed to products plus the anticipated convection and radiation heat losses divided by the weight of water to be removed per hour. If a line is now drawn on the humidity chart from the temperature and humidity conditions of the hot inlet air to the outlet dry-bulb temperature ordinate parallel with the line joining the arbitrary point to the appropriate heat loss factor value on the scale, the exit air humidity can be read off the humidity ordinate, thus establishing the outlet air conditions. The difference between the inlet and the outlet air humidities is now known and hence the quantity of air required can be calculated. The dryer manufacturer will know fairly accurately from experience the heat losses from different sized plants, and also the size of his standard cyclone which will deal with the calculated quantity of air effectively, and when taken in conjunction with the length of trunking required to suit the cyclone this will give an adequate drying time for the drying operation. Based on his experience, pilot-scale tests which yield the product properly dried can be used directly to arrive at the size of a commercial plant without application of any additional scale-up factor.

By reason of the small particle size, the high gas mass velocities and turbulence, and the considerable initial temperature differences between material and gas at the point of initial contact, conditions in a flash dryer are conducive to optimum drying conditions and minimum drying times. The material will remain at the wet-bulb temperature of the gas during the major part of the drying cycle due to the high rate of moisture evaporation and will at no time exceed the outlet gas temperature. Where materials have an exceedingly high moisture content – upwards of 60% on the wet-weight basis – there may be a limitation to the effectiveness of a flash dryer due to the excessive air velocity and hence reduced residence time as a result of the release of large quantities of moisture vapour. In such a case, it is usual to recycle some of the dry product in order to reduce the average moisture content of the feedstock. Some designs of flash dryer incorporate expansion sections in the main duct in order to increase the residence time as moisture is liberated. Where the initial moisture content is high, thermal efficiencies will be invariably high, eliminating the necessity for any gas recycling; whether gas recycling is adopted or not, the exhaust fan will of course require to handle the total gas exhausted together with the released moisture vapour at the temperature on leaving the main cyclone. Unless there is a pressure fan at the inlet end, the exhaust fan will also have to overcome the total pressure losses of the system including venturi, ductwork and cyclone, which will usually be quite considerable. Where the fan is located at the inlet end, it is usual to install it after the heater, in order to reduce the power requirements, cold starting being carried out against a partially closed damper to reduce the mass of gas moved.

Pneumatic dryers are probably the most suitable form of atmospheric pressure continuous direct dryers with which to dry materials having their moisture content in the form of recoverable solvents. The humid exhaust air leaving the cyclone and dust recovery equipment will be passed to a suitable condenser and the solvent recovered. A steam heater unit will usually be used as the source of heat, in which event the exhaust air, stripped of as much solvent as possible, may be recirculated through the heater and the dryer operated as a closed system. Where inflammable or explosive solvents are present, special care must be taken in earthing all parts of the dryer. In such cases and where materials which form explosive mixtures with air are dried, adequate provision of explosion panels or discs must be made. In extremely dangerous cases, a neutral gas may be used on the closed system principle employing a suitable cooling system for moisture removal from the gas.

Pneumatic Dryer Systems

The A.P.V.–Kestner T–V Dryer

The A. P. V.–Kestner Thermo–Venturi dryer is one of the simplest designs of flash dryers and is illustrated diagrammatically in Fig. 47. Essentially it consists of a vertical drying column having a convergent-divergent section

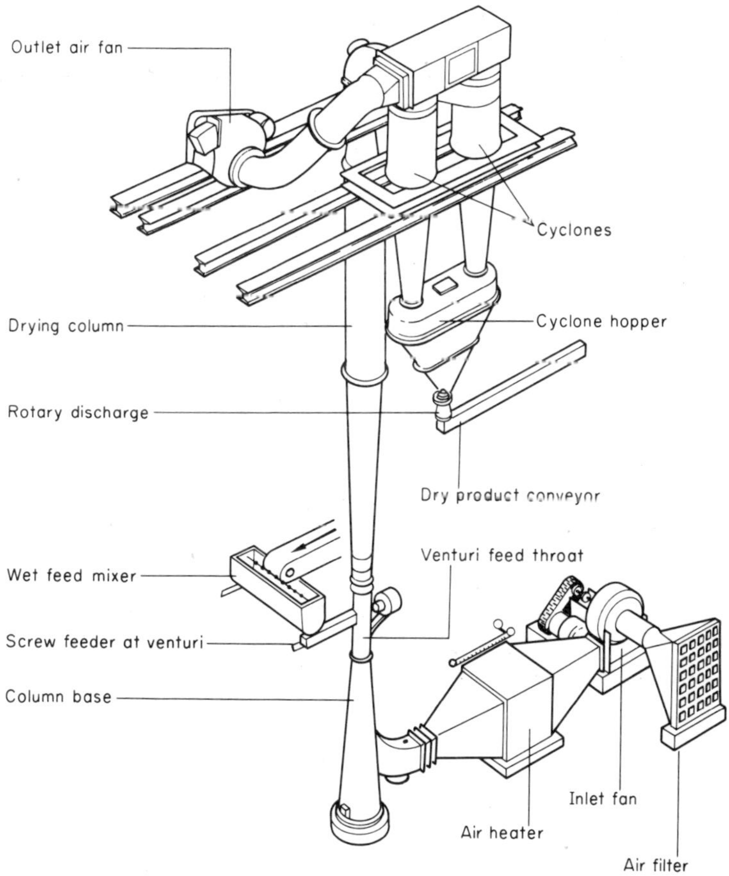

Fig. 47. A.P.V.–Kestner T–V dryer. *Courtesy A.P.V. – Kestner Evaporator & Engineering Co. Ltd.*

at its base forming a venturi into which the feedstock is fed and in which it becomes air-borne. Pilot-scale tests establish the optimum gas velocity within the venturi section and the diameter of the straight duct section so that the largest particles of feedstock are almost in a state of aerodynamic balance, there being only a slow upward movement of these particles in the gas stream, thus ensuring maximum residence time. Tests also establish the requisite length of straight section to attain uniform final residual moisture content. Gas velocities are increased in a convergent section to those required at entry to the cyclone for optimum solids-gas separation in the latter. Where necessary some dried product from the cyclone may be re-introduced into the system through a back-mixer with the wet feedstock.

It is usual in this system to provide a fan ahead of the dryer and at the exhaust from the cyclone, the latter passing exhausted gas to a fabric dust-collecting unit or a wet scrubber if necessary. Adjustment of dampers adjacent to these fans will ensure balanced pressure conditions with a negative pressure at the venturi feed point if desired. To meet the case where available headroom is restricted, this company has developed two modified designs, namely, their Contra-flow dryer and their S–T–V dryer. These designs, particularly the latter, make it possible to extend the residence time so as to deal with less easily dried material without demanding excessive column heights. These two dryers are illustrated diagrammatically in Fig. 48.

In the contra-flow dryer, a double column is employed with an expansion chamber connecting the two. Feedstock enters the venturi at the base of the inner column in which the initial stages of drying occur and then passes entrained in the gas via the expansion chamber to the outer column, down which it passes to the cyclone separator. Additional residence time is provided in the expansion chamber and final drying takes place in the annular space between the two columns. Heat losses from the inner column are utilised in the outer column which with the expansion chamber and cyclone above requires to be insulated. The S–T–V dryer has been designed to deal particularly with materials whose moisture content is only to a lesser extent surface moisture and where the diffusion of moisture through the material particles is largely controlling. Increased residence time is required to complete drying and this is achieved by designing the drying column as a series of vertical columns arranged in a serpentine manner, the pneumatically suspended material passing upwards and downwards through these columns in series to the cyclone separator. The drying effect in the upward and downward legs of the system is, of course, not similar, as the solid's velocity in the downward legs is about twice that in the upward legs, but an overall considerable increase in residence time as compared with a single vertical column of the same height is achieved. The

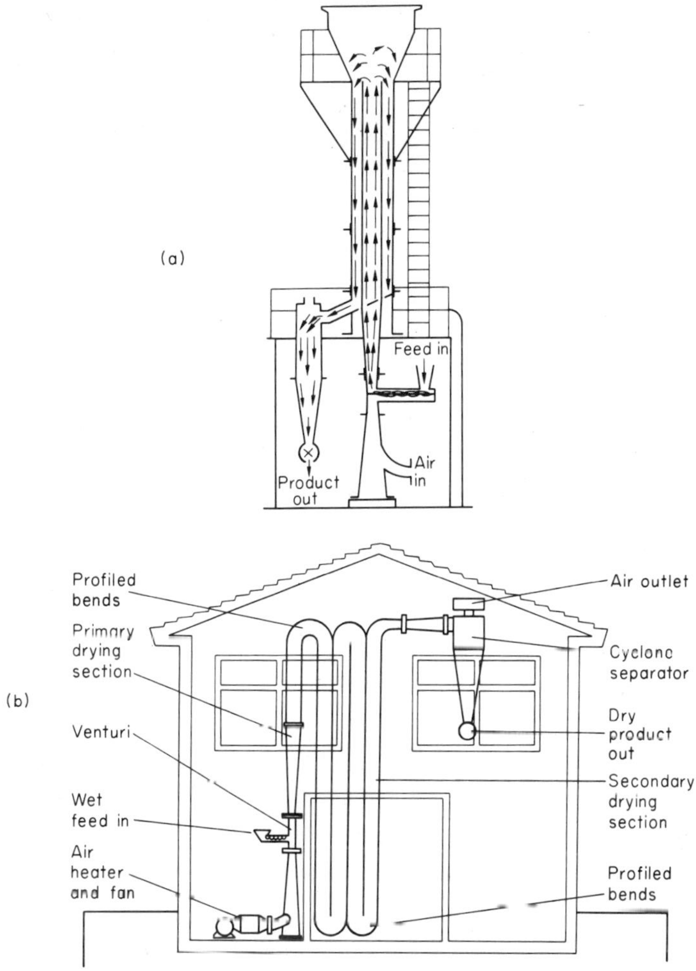

Fig. 48. A.P.V.–Kestner Contra-flow and S–T–V dryer.

introduction of curved baffle plates in the bottom bends has been found to overcome a tendency towards accumulation of solids at these points.

The Rietz Air-Lift Dryer

This dryer, illustrated diagrammatically in Fig. 49, is designed in such a manner that material entrained pneumatically in an initial high-velocity section at the base of the drying column passes through one or more expansion chambers in series integral with the column. In these, gas velocities are alternately reduced and increased; heavier particles remain

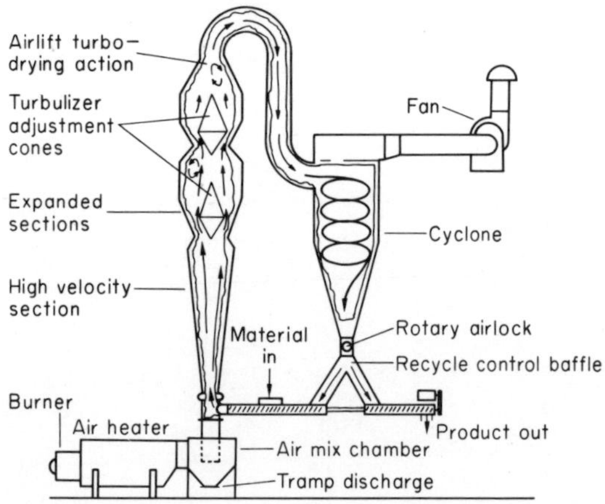

Fig. 49. Rietz air-lift dryer.

suspended until their weight is reduced as drying proceeds and are then carried onwards in the gas stream to the next expansion section where the process is repeated. Turbulent conditions are maintained in the expansion sections – which are of double conical shape – by adjustment of turbuliser cones located at the entry to the expansion sections. Fully dried material is carried forward from the last chamber to the cyclone recovery system. Material recirculation is possible in the case of materials very resistant to drying or where back-mixing is necessary with very wet feedstock.

The Proctor–Mark P. B. Dispersion Dryer

This operates on a somewhat similar principle to the Rietz air-lift dryer but has an optional refinement permitting the introduction of secondary supplies of hot gases tangentially at the expansion sections in order to induce helicoidal trajections to the gases within the expansion sections, thus still further enhancing the residence time of larger and heavier particles.

The Proctor–Mark P. Type and J. S. Type Pneumatic Dryers

The P-type of dryer is in most respects conventional in design and operation but secures considerable compactness whilst still maintaining high efficiencies. It is illustrated diagrammatically in Fig. 50.

It will be seen that the feedstock is introduced into a venturi – usually horizontally placed slightly clear of the floor level – of effective design to maintain an internal reduced pressure to assist material feeding. Drying medium and feedstock pass through the drying duct to a high efficiency cyclone of relatively large size in which drying is completed and located

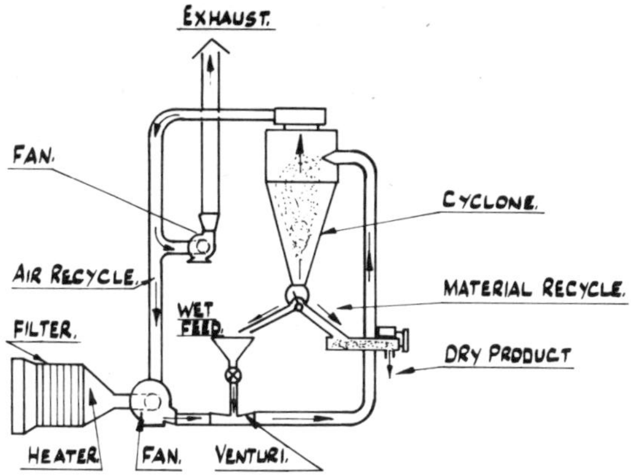

Fig. 50. Proctor–Mark P–type pneumatic dryer.

slightly to one side of the venturi feed point. Bifurcation of the discharge from the cyclone leads recycle stock to the original feed point, the remainder passing onwards as dried product. The recycle ratio is conveniently controlled by a proportionating valve damper within the base of the cyclone. The main gas stream is usually provided by a fan ahead of the material feed point and any air recirculated is introduced into the eye of this main fan. A separate exhaust fan withdraws gases and moisture vapour from the connecting duct.

The J. S. Dispersion or pneumatic dryer, shown diagrammatically in Fig. 51, is a less conventional design and is suitable for materials which are

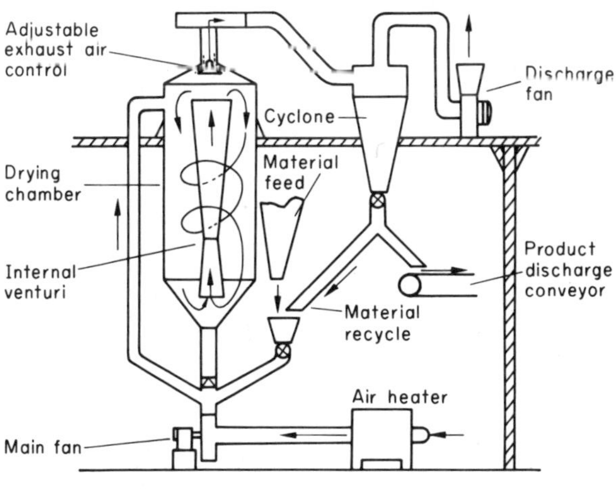

Fig. 51. Proctor–Mark J.S dispersion dryer.

heat-sensitive and which may have a wide particle size range. The feedstock entrained in the gas stream is injected at high velocity upwards into a free venturi supported within a vertical cylindrical drying chamber thus producing a zone of suction at the venturi entrance. Fine, easily dried material is ejected from the venturi outlet to the exhaust point of the chamber, whilst heavier less easily dried material is subjected to internal recirculation down the chamber and upwards within the venturi. The recirculation ratio is controllable whilst the drying time for heavier particles can be still further increased by tangential introduction of hot gases in the upper part of the drying chamber, which imparts a helicoidal trajectory to the gases and suspended solids. Exhaust gases from the chamber discharge pass to a cyclone, whence dry product is extracted, a proportion of which may be back-mixed with initial feedstock.

Buttner-Rosin Flash Dryer

These dryers are conventional pneumatic dryers with one or two design refinements particularly suited to materials of wide particle size. Such a dryer is illustrated diagrammatically in Fig. 52.

Drying gases with entrained solids, after passing through a vertical drying column, pass through a specially designed pneumatic sifter and, if required, a classifier, the separated heavier and larger particles being led either to a feed point adjacent to the base of the column or to a disintegrator connected to the column base. Large particles are thus subjected to recirculation with prior disintegration if required. The disintegrator

Table 24. Performance data on Buttner-Rosin pneumatic dryer

	Metallic stearate	*Starch*	*Adipic acid*	*Fibre*	*Coal filter cake*
Method of feed	Sling	Sling	Screw	Distributor	Distributor
Material size	Fine	Fine	−30 mesh	$-\frac{1}{4}$ in	−30 mesh
Product rate, lb/h	280	13,236	10,000	2,610	67,200
Initial moisture, % w/w basis	40	34	10	62·4	32
Final moisture, % w/w basis	0·5	13	0·2	10	6
Air inlet temperature, °F	284	302	320	752	1,292
Air outlet temperature, °F	130	122	149	230	212
Method of heating	Steam	Steam	Steam	Oil	P.F.
Heat consumption Btu/lb/ water evaporated	2,170	1,825	2,400	1,720	1,590
Air recirculation,	No	No	No	No	Yes
Material recirculation,	Yes	No	Yes	Yes	Yes
Fan capacity, Std ft^3/min	1,440	26,500	9,500	12,500	27,000
Installed fan hp	15	220	65	60	250
Product exit temperature, °F	104	95	120	140	158

Courtesy Rosin Engineering Ltd.

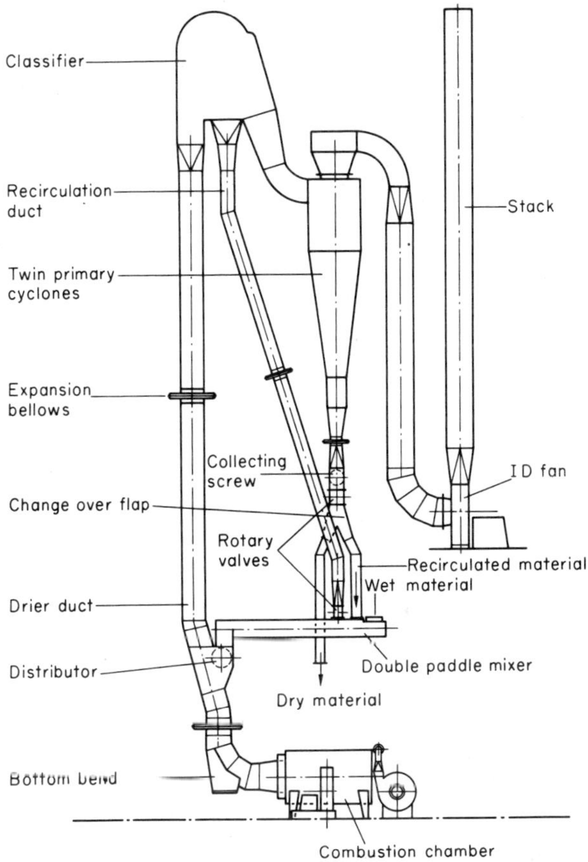

Fig. 52. Buttner Rosin flash dryer.

can itself be supplied with hot gases to further assist in the drying. An alternative arrangement is composed of a two-stage dryer, the larger particles from the classifier passing to the second-stage dryer.

The Berk Ring Dryer

This dryer, manufactured by Pennsalt Ltd., departs materially from conventional pneumatic dryer design. As will be seen from Fig. 53, the drying duct is an endless ring duct through which the feedstock is circulated pneumatically by the hot gases. Humid exhaust air is continuously extracted and continuously replaced by hot inlet gases at a common manifold of circular design, the exhaust air and dried product leaving at the centre. Centrifugal force within the manifold throws larger and heavier particles

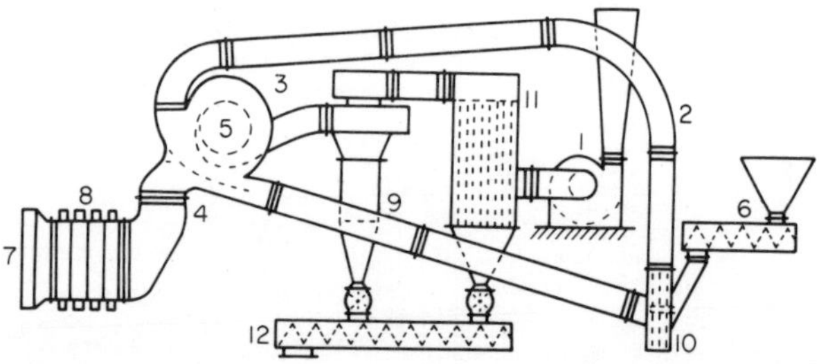

Fig. 53. Pennsalt–Berk ring dryer.

outwards which are then picked up by the gas stream and recirculated in the system to give increased residence time by means of a built-in high velocity hot gas jet.

A disintegrator can be incorporated in the ring system and subjected to the drying action of the hot gases. Large particles in the system can be reduced to the final required size and absolute uniformity of drying is thereby assured. In the absence of a disintegrator within the circuit, the larger particles continue to recirculate until through loss of moisture their particle weight is sufficiently reduced for them to leave the system at the manifold centre with the main stream of dry product.

Table 25. Performance data on pneumatic dryers*

	Metal[1] stearates	*Spent[2] grains*	*Sewage[3] sludge*	*Starches*	*Polystyrene beads*
Method of feed	Belt feeder rotary valve	Back-mixer rotary valve	Vibratory feeder rotary valve	Cascading rotary valve screen	Vibratory feeder rotary valve
Product rate, lb/h	240	1,120	4,300	5,000	1,000
Initial moisture, % w/w basis	55	80	45	35	2·0
Final moisture, % w/w basis	1	5	12	10	0·2
Air inlet temperature, °F	250	500	600	300	175
Air outlet temperature, °F	150	170	170	130	115
Method of heating	Steam	Gas	Oil	Steam	Steam
Heat consumption, Btu/lb water evaporated	2,900	1,800	1,750	2,000	5,000
Air recirculation,	No	No	No	No	No
Material recirculation,	Yes	Yes	Yes	No	No
Material of construction,	S.S.	M.S.G.	M.S.	M.S.G.	S.S.
Fan capacity, std ft^3/min	3,750	16,500	8,250	15,000	900
Installed fan hp	20	75	60	60	7·5

*Items 1, 2 and 3 refer to ring dryer applications.

Courtesy Pennsalt Ltd.

The Raymond Flash Dryer

This dryer, manufactured by International Combustion Products Ltd., in Great Britain, and by Combustion Engineering Co. in America, follows conventional form whether with or without a disintegrator in the circuit. The vertical drying column is usually followed by a cyclone separator of ample dimensions which increases the residence time. When furnished with a disintegrator, the feedstock and hot gases are usually both fed into the disintegrator so that initial drying occurs within it. With pasty and very wet materials, dry product is recirculated via a back-mixer. In this system, the dryer is maintained under suction produced by the exhaust fan located after the dust recovery equipment.

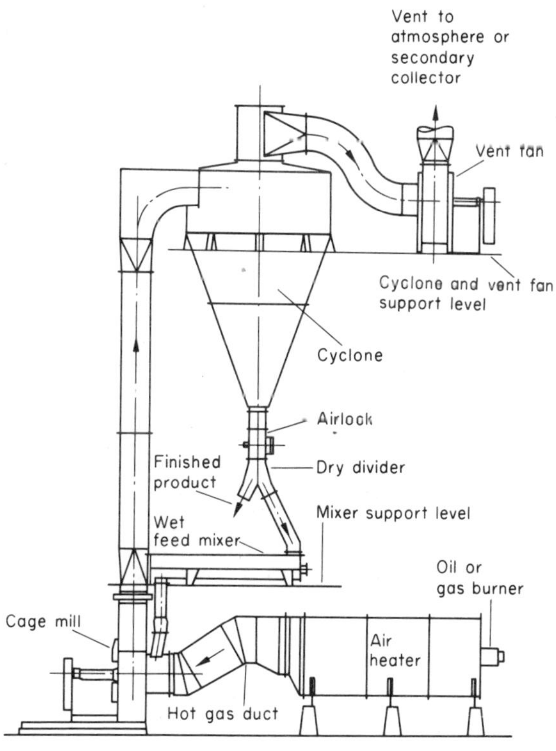

Fig. 54. Raymond flash dryer. *Courtesy International Combustion Products Ltd.*

Table 26. Performance data on Raymond flash dryer

	Fine mineral	*Spent grain*	*Organic chemical*	*Chicken droppings*	*Fine coal filter cake*
Method of feed	Pump	Belt	Screw	Pump	Screw
Material size	−100	—	−30	—	−30
Product rate, lb/h	27,000	9,000	900	2,300	2,000
Initial moisture content, % w/w basis	25	60	37	70	30
Final moisture content, % w/w basis	Nil	12	3	12	8·5
Air inlet temperature, °F	1,200	1,200	450	1,300	1,200
Air outlet temperature, °F	200/300	200/300	200/300	200/300	200/300
Method of heating	Direct oil	Direct oil	Direct oil	Direct oil	Direct oil
Heat consumption, Btu/lb water evaporated	$1{\cdot}6 \times 10^3$	$1{\cdot}9 \times 10^3$	$3{\cdot}1 \times 10^3$	$1{\cdot}9 \times 10^3$	$1{\cdot}4 \times 10^3$
Air recirculation,	No	No	No	No	No
Material recirculation,	Yes	Yes	No	Yes	No
Material of construction,	M.S.	M.S./S.S.	M.S.	M.S.	M.S.
Fan capacity	18,000	22,000	4,300	8,500	1,500
Installed fan hp	110	180	30	50	10
Product exit temperature, °F	200	—	200	—	135

Courtesy International Combustion Products Ltd.

The Atritor–Dryer Pulverisor

The Atritor, manufactured by Albert Herbert Ltd., Coventry, was originally designed to pulverise and dry coal used in the firing of cement kilns but has since found very wide application in industry. The pulverising zones, which are the main feature of this dryer, are of unique design and pulverise the feedstock by pure attrition. It is normal practice to introduce both feedstock – back-mixed with some dry product – and combustion gases directly into the Atritor-Pulverisor thus accelerating the drying effect in the early stage. The hot gases carry the feedstock after preliminary disintegration in the hammer mill section of the Atritor to the second, or attrition, zone where the material is repetitively carried forward towards the centre and outwards to the periphery until the final degree of pulverisation has been attained. The intense turbulence set up enhances the preliminary drying effect before entering the duct leading to the solids – air cyclone separator in both of which drying is completed. Maximum thermal efficiencies are obtained with an inlet gas temperature of 950°F and exit gas temperature of 180°F when oil firing. Evaporation rates of 12 lb/lb of oil are normal under these conditions; some large users report evaporation figures of 13 and 14 to 1, and 400 lb of moisture for every 1,000 ft^3/min of gases delivered by the fan.

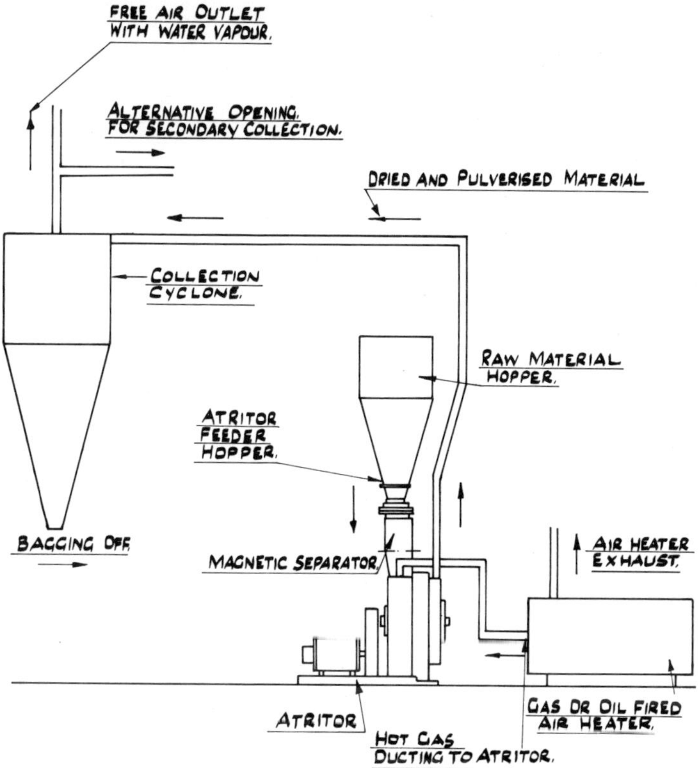

Fig. 55. Alfred Herbert Atritor dryer.

Table 27. Typical performance of Atritor dryer

	Limestone	*Clay*	*Chalk*	*Filter cakes*
Product size and mesh, B.S.S.	50%–100 mesh to 99%–200 mesh	$-\frac{1}{8}$ in to 98% −200 mesh	$-\frac{1}{8}$ in to 99% −300 mesh	$-\frac{1}{8}$ in to 99% −300 mesh
Initial moisture, % w/w basis	1–10%	5–20%	5–20%	20–80%
Final moisture, % w/w basis	Under 1%	25–10%	Under 1%	1–10% as required
Inlet temperature, °F	400 to 700	400 to 950	Up to 950	Up to 950
Outlet temperature, °F	150 to 175	120 to 180	180 to 220	150 to 220

Courtesy Alfred Herbert Ltd.

Buttner Jet Dryer

The design of this dryer, manufactured by Buttner Werke in Germany and Buell Ltd. in Great Britain, departs materially from conventional flash dryer designs but can be considered to operate essentially as a pneumatic dryer. A somewhat similar dryer, known as the Jetstream Dryer, is manufactured in the U.S.A. by Engineers Associated Inc. of California. This form of dryer is illustrated diagrammatically in Fig. 56.

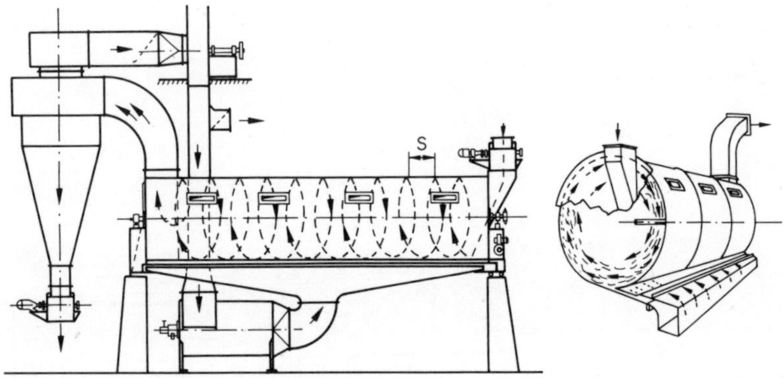

Fig. 56. Buttner–Buell jet dryer.

It was originally developed for the air-borne drying of wood chips in the particle board industry but has been successfully applied to other materials of fairly large particle size of low density. Essentially it consists of a stationary horizontal drying cylinder into which feedstock is introduced at one end and discharged pneumatically at the other end to a cyclone separator. Heated gases are introduced tangentially at high velocity through jets extending the whole length of the drying cylinder and fitted to a plenum below the cylinder. A spiral peripheral movement of the gases takes place which maintains the material subjected to drying in suspension, this being assisted by a rotating agitator revolving in the same direction. The angle of entry of gases at the jets or nozzles can be manually adjusted by external handwheels actuating movable vanes, the pitch of the gas spiral being thus controlled at different points down the length of the dryer. This, in turn, permits adjustment to the retention time of the product at different stages in its travel. Provision can be made for recirculating the products of combustion on the heater and dryer if required.

Table 28. Performance data on Buttner-Buell jet dryer

	Wood chips	*Wood chips*
Method of feed	Airlock	Airlock
Material size	$\frac{3}{8}$ in × $\frac{1}{32}$ in × $\frac{1}{4}$ in	$\frac{1}{2}$ in × $\frac{1}{64}$ in × $\frac{1}{8}$ in
Product rate, lb/h	12,700	4,030
Initial moisture, % w/w basis	55%	40%
Final moisture, % w/w basis	4%	7%
Air inlet temperature, °F	662	572
Air outlet temperature, °F	284	284
Method of heating	Oil	Oil
Heat consumption, Btu/lb water evaporated	1,540	1,640
Air recirculation,	Yes	Yes
Material recirculation,	No	No
Material of construction,	M.S.	M.S.
Fan capacity, Std ft^3/min	60,000	13,000
Installed fan hp	160	40
Product exit temperature, °F	190	190

Courtesy Buttner-Werke.

CHAPTER IX

Fluid Bed Dryers

Fluidisation

When a vertically rising stream of gas is introduced at the base of a bed or column of particulate solids there exists a velocity of this air stream at which the solids bed will be expanded to a point at which the particles will move within the bed. Each particle is surrounded by the gas and is suspended and agitated by the gas stream, the relative velocity of gas to solid being high although the gas velocity itself may be reasonably low. This stage is known as the onset of fluidisation or a state of incipient fluidisation. Further increase in the velocity of the gas stream will greatly increase the expansion of the solids bed, there is considerable relative movement of solid particles and the dispersed particles in the gas stream acquire many of the properties of a liquid, having similar flow characteristics and exerting hydrostatic head. There will be different degrees of fluidisation depending upon the relative velocities of gas stream and solids ranging from incipient, through dense, to dilute fluidisation, and finally, if carried to extremes, total suspension of solid particles in the gas stream, at which point fluidisation has ceased and pneumatic conveyance has occurred. Dense fluidisation, synonymous with medium turbulence, creates partial dispersion of the solids whilst dilute fluidisation indicates heavy turbulence and an approach of the particle bed to the disperse phase.

If a bed of moist material is fluidised by a hot gas stream, conditions will be almost ideal for drying. Intimate contact between gas and solids at a desirable relative velocity will promote high heat transfer rates by a combination of convection and conduction. Such high heat transfer rates, especially when free moisture is present, will result in almost instantaneous evaporation, with consequent rapid fall in gas temperature at the gas stream entry area. Extremely efficient mixing of the solid particles occurs, giving a high degree of uniformity of drying and with reduction in moisture content the material is in contact for a fraction of time with gases the temperature of which has been materially reduced. It is evident that the technique of fluidisation can be applied either to a batch of material or, if production considerations warrant it, to a continuous flow of material. In either case, however, the gas stream velocity must be controlled to yield optimum conditions for drying having regard to the particle size and its density. This air velocity will lie at some point between that required for incipient fluidisation and the dilute stage, but certainly below the point of major entrainment, as this would result in the necessity for considerable and expensive solids recovery systems. Furthermore, the gas stream velocity should not be such as to set up a condition of bubbling or "slugging"; that is, a condition where the concentration of solids is not uniform throughout the bed. In the case of a boiling bed, part of the gas stream passes in the form of bubbles the size of which may be not much larger than the particle size but very small relative to the dryer vessel or container. When these bubbles attain considerable size relative to the particle size and yet small relative to the size of the container, the bed is said to be "bubbling" and in beds of relatively low cross-sectional size, as in a column, the bubbles may occupy the whole of the cross-section and result in "slugging". The latter is a particularly undesirable condition and for operational reasons should be avoided. During the upward movement of the slugs through the fluidised bed, particles separate from the slugs, falling back until the slugs collapse, new slugs being formed in the lower regions of the bed. Steady operating conditions are impossible in consequence. Plate XXV illustrates types of fluidised bed conditions.

Fluidised bed dryers are designed to introduce the hot gas stream at the base of the bed, whether in a container or a column, through a dispersion plate generally of substantially the same base area as that of the bed enclosure. When the gas enters through a single opening of restricted area in relation to that of the enclosure, what is known as a "spouting" bed is produced in which the gas stream carries the solid particles to the surface of the bed in the form of a core of dilute phase, the material around this core taking a downward movement. Such a condition is generally avoided in fluidised bed dryers through the use of correctly designed dispersion plates. In a properly designed and operated fluid bed dryer the degree of fluidisation is such as to produce full agitation of the solid particles, which are to a small extent ejected from the surface of the fluidised bed resulting in the absence of any apparent discrete interface. Each particle being surrounded by gas, very little attrition of particles and abrasion of the container is usually experienced, there is free movement of the individual particles and mixing is substantially instantaneous. Temperature throughout the bed becomes extremely uniform and it is not unusual for the material

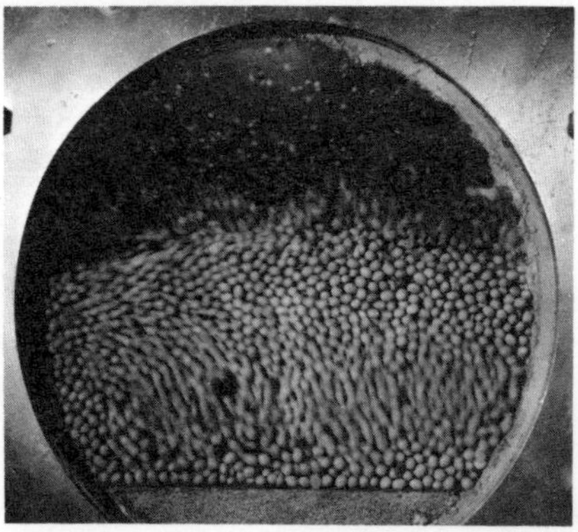

(a) Incipient.

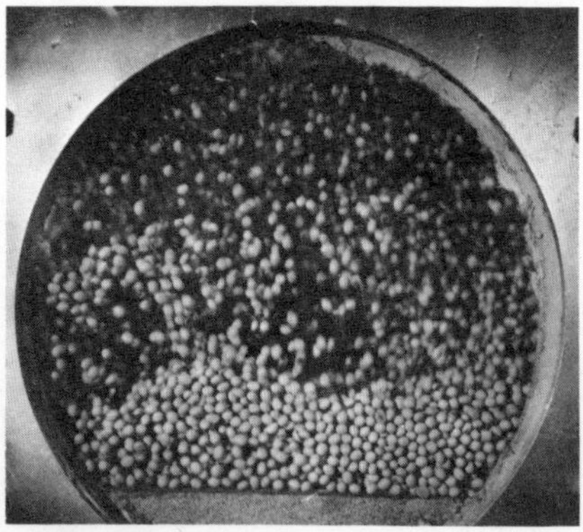

(b) Dense.

(c) Dilute.

Plate XXV. Degrees of fluidisation. *Courtesy Link-Belt S.A.*

and the gases leaving the bed of a continuous fluid bed dryer to be at the same temperature. Heat-transfer rates are of a very high order, frequently attaining 50 Btu/h/ft^2/deg F.

As the superficial gas velocity supplied to a static bed of granules increases, the pressure drop in the gas increases, the bed depth meanwhile tending to expand. This pressure drop persists up to the point of onset of fluidisation. As the gas stream velocity is increased beyond the stage when fluidisation is established, the pressure drop remains substantially constant except in so far as it is affected by the resistance of the distributor plate. This is illustrated in Fig. 57 where pressure drop is plotted against gas velocity. Curve (*a*) represents the pressure drop through the distributor plate alone, curve (*b*) that of the granule bed and curve (*c*) the combined resistance of the distributor plate and bed. It will be seen that the pressure

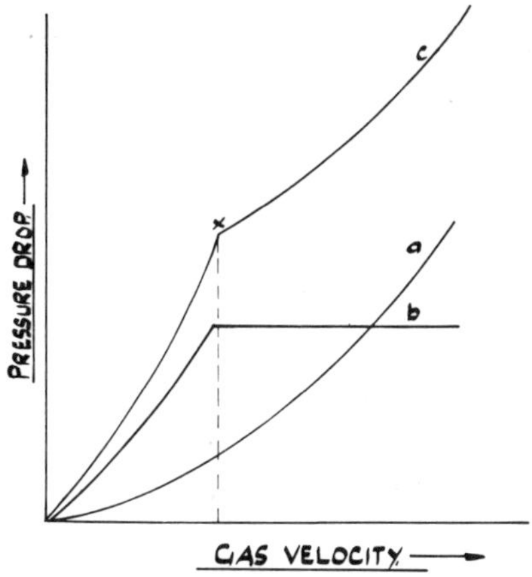

Fig. 57. Pressure drop through fluidised beds.

loss curve (*c*) shows a sharp change of direction at the point of incipient fluidisation (*x*). Depending upon the sieve analysis and density of the granular material and the type of distributor plate, the pressure loss from all causes can be considerable, and it is usual to operate fluid bed dryers with the gas stream pressure fan at the inlet to the system rather than at the outlet, a position favoured with pneumatic dryers where pressure losses may not be so high depending on the design and type of dust recovery equipment employed.

Not all materials are in such a physical form that they can be fluidised and before considering the possibility of using a fluidised bed dryer it is necessary to establish that the material as produced or admixed with a proportion of dry product lends itself to this form of treatment. It is customary to carry out simple preliminary tests with a glass column of, say, 3 in or 4 in diameter, fitted with a suitable disperser plate or grid in the form of a wire mesh through which air is passed, a bed of the material being supported on the wire-mesh grid. The onset of fluidisation can usually be observed visually, but if not the flattening-out of the pressure drop curve at increasing air stream velocity will indicate that fluidisation is taking place. It will also be possible to ascertain the degree of material entrainment to be anticipated above fluidising velocities and if this is excessive it may mean consideration of an alternative type of dryer such as a pneumatic dryer.

Batch Fluid Bed Dryers

In batch fluidised drying, the wet material is introduced into the dryer as a batch of material in a container provided with a suitable distributor plate, usually a perforated or wire-mesh metal supporting screen, or one composed of a woven synthetic fibre material, and is subjected to a stream of heated air at a selected temperature. When the material is sufficiently dry it is removed from the dryer. The temperature of the air stream at the inlet may be maintained constant throughout the drying operation, in which case, with low final moisture content, the material itself will probably attain the same temperature; or the initial temperature of the air stream may be reduced as drying proceeds to a level at which the dried product is known not to suffer any deterioration. The latter method is more economic. In the early stages of drying, when the rate of drying is usually at its highest, the rate of moisture evaporation depends upon the rate of heat input, the material remaining substantially at the wet-bulb temperature. High initial air temperatures will therefore reduce drying times and increase the number of charges which the dryer will handle. As drying attains the critical moisture content, the bed temperature will begin to rise and should be limited to a safe temperature, either by thermostatic control of the inlet temperature actuated by a heat-sensitive element in the bed or in the air stream leaving the bed, or by means of a programme-controlled thermostat the operation of which has previously been determined by experience on the particular material. In the exceptional case of the wet material being more heat-sensitive than the dry material, it would not, of course, be possible to have higher initial air temperatures than the air temperature towards the end of the drying cycle. When drying material to a predetermined figure other than dryness, as for instance to the equilibrium moisture content value, the temperature of the air will have to be selected to give the desired result.

A typical batch fluidised bed dryer is shown in Plate XXVI and diagrammatically in Fig. 58. Such dryers will usually be available with material capacities ranging from $2\frac{1}{2}$ to 15 or 20 ft^3, and weight loadings from 50 to 500 lb. They are ideally suited to the batch drying of granular, powdery or crystalline products in the pharmaceutical, fine chemical, dye and pigment and plastics industries. They are progressively displacing forced convection tray-drying cabinets and ovens for small and medium productions in view of lower labour requirements, compactness and reduction in drying times. It is always, of course, necessary to establish that the wet product can be fluidised and, as most batch fluid bed dryer manufacturers possess demonstration test units, this presents no difficulty. Due to the greatly accelerated drying rate possible, one such batch dryer can usually replace a number of tray dryers, with consequent saving in factory floor space. Furthermore, the required production can usually be obtained by operating the dryer during a normal working day whereas a tray dryer might necessitate overnight drying. Instead of charging and withdrawing a large number of trays, all the material in a production batch can often be dealt with in a single or a few fluid-bed containers. Fluid-bed dryers can usually be maintained in

Plate XXVI. Batch fluidised bed dryer. *Courtesy Calmic Engineering Co. Ltd.*

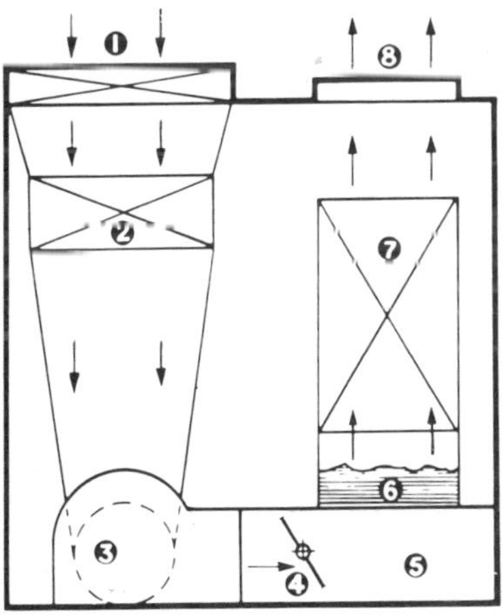

1 Fresh Air Inlet Filter.
2 Steam or Electric Heater.
3 Air Circulating Fan.
4 Air Volume Control Damper.
5 Air Diffuser.
6 Tub containing material.
7 Nylon Filter Bags.
8 Air Exhaust.

Fig. 58. Batch fluid bed dryer—pressure type. *Courtesy Calmic Engineering Co. Ltd.*

a very high state of cleanliness, and if of the total rejection type can deal with more than one product in consequence. They do, however, require careful supervision in their operation if optimum results are to be obtained and it could be argued that a more intelligent class of operative is required than when using tray dryers which can be left wholly unattended during the drying cycle.

As it is the movement of the air through a bed of material in the container which causes fluidisation, this can be achieved either by blowing the air through the bed as shown in Fig. 58 or by sucking the air through the bed. Each system has its advantages and disadvantages. The suction type lends itself to a more compact design, as the exhaust fan can be mounted immediately above the material container and filter bags and only one seal is required between the material container and the base of the filter bag assembly. Access doors in both types must make a good seal. However, with the suction type, the exhaust fan must handle all that passes the filter bags, both dust and vapour, and this may sometimes add to the explosion hazard. There is also rather less flexibility in operation on some materials than with the pressure type. In the latter, once the drying air has passed through the material and the filter bags, it has a clear passage to atmosphere; the fan only handles clean air free from volatile solvents and explosion risks are lower. The general design of the drying chamber can be more open and anti-explosion precautions can be of simpler design. However, as the whole of the drying compartment is under pressure, both the top and bottom flanges of the material container must make good seals at all times and any leakage, being outwards, can be the cause of inconvenience and possibly loss of product. As the fan will have to be located in a side chamber, the pressure type may take up more floor space than the suction type. Some batch type fluid bed dryers are designed for partial recirculation of air, but this design is not favoured where there is danger of solvent build-up in the air stream. As organic solvents are frequently present in pharmaceutical products which have to be dried, this is a matter of considerable importance and for this reason batch fluid bed dryers have built-in earthing features.

The violent agitation of the material in its container by the air stream and the passage of air and dust through the filter bags leads to the generation of static electricity, varying in quantity in different materials and the degree of dryness of the product. The build-up of static electricity can be considerable and in the absence of adequate precautions can lead to serious explosion. The static charge build-up in the material does not usually present a difficult problem as the container, being usually of one or another quality of steel can be effectively earthed. If, however, the container is fabricated in a non-conductor such as Perspex, special precautions have to be covered by the design of assembly. With metal containers, as there is usually a rubber or some other non-conducting gasket at the seals, it is usual to incorporate earthing leaves which make a good contact between the container itself and the dryer structure, which in turn is earthed. These earthing contacts should be quite automatic in their function and should not be left to an operator to adjust. The static build-up in the filter bags, whether of cotton or synthetic fibres, presents a more difficult problem. Synthetics having much lower conductivities than cotton lead to considerable static build-up which may cause sparking. Special provision has to be made to keep filter bags in a discharged condition and it is now normal practice to employ a fabric in which copper or stainless steel wires are woven into the fabric itself and to space these conductors sufficiently close together that the amount of charge carried by each is small. The filter bags and the wires woven into them make earthing contacts with the metal clamps or clips at each end. Because of the danger of these metal threads breaking with constant use and washing of filter bags, the modern tendency is to use fabric with interwoven graphite-coated threads which overcome the objections to metal conductor wires. Where explosive solvents are present or when the material itself can form explosive mixtures with air, it is important that the exhaust fan on the suction-type fluid bed dryer should be of spark-proof design and that any belt drive should be of an anti-static type. None of these precautions is completely foolproof and where the components of an explosive mixture are present the explosion hazard must be taken into consideration. It is not unusual to incorporate explosion discs or panels leading to atmosphere, and in extreme cases to house the dryers in a suitable separate area or blast-proof enclosure. The whole question of the susceptibility of a material to creating an explosion hazard under fluid bed drying conditions should be thoroughly discussed by a potential user with the dryer manufacturer, as the latter will have examined this whole problem with the factory safety authorities and will have designed his equipment to meet most possibilities.

Typical performance data for a batch fluid bed dryer are given in Table 29.

Continuous Fluid Bed Dryers

General

As stated earlier, fluidised solids exhibit many of the properties of a liquid. They will seek their own level, show similar flow characteristics to those of a liquid, flowing over weirs or in pipes, and exert hydrostatic head. Thus, if a continuous flow of material is introduced into a suitably designed

Table 29. Batch fluid-bed dryer performance data

	Ammonium bromide	*Lactose base granules*	*Pharma-ceutical crystals*	*Liver residue*	*Weed killer*
Holding capacity, lb wet product	100	104	160	280	250
Bulk density, dry, lb/ft^3	75	30	20	30	35
Initial moisture, % w/w basis	6	10	65	50	20–25
Final moisture, % w/w basis	1	2	0·4	5·0	1·0
Final drying temperature, °F	212	158	248	140	140
Drying time, min	20	90	120	75	210
Fan capacity, ft^3/min at 11 in w.g.	750	1,500	3,000	4,000	3,000
Fan hp	5	10	20	25	20
Evaporation rate, lb H_2O/h	15	5·7	52	100	17

Courtesy Calmic Engineering Co. Ltd.

container, such as a cylindrical column, provided with an overflow discharge pipe or weir and a supporting perforated or grid base through which the fluidising gas is introduced, the fluidised solids can be continuously withdrawn at a controlled rate such as to produce a fluidised bed within the column of desired height. The retention time of the solids within the column can thus be controlled and its hydrostatic head can be maintained constant. The solids in the bed are intimately mixed during their period of retention and will be in the most intimate contact with the fluidising gases. The degree of mixing in most cases is so high that the properties of the material hold-up in the bed are substantially the same as those of the material discharged from the bed. If the feedstock is initially moist and the fluidising gases hot, such an apparatus constitutes a highly efficient continuous dryer. Conversely, if the solids are initially hot and dry and the fluidising medium is cold, the apparatus will become an efficient continuous cooler.

The advantages inherent in such a system, such as the high heat-transfer rates obtainable, the extremely rapid reduction of the fluidising medium temperature at the point of introduction and the uniformity of temperature within the bed, have already been stressed. However, certain limitations and possible disadvantages also exist and should be noted. The almost complete uniformity of bed temperature, except in immediate proximity to the fluidising base, although of advantage in the case of heat-sensitive materials, greatly reduces any desirable counter-current thermal interchange effect, as the temperature of the solids leaving the bed is substantially that of the gases leaving the bed, and this, taken together with the intimate physical mixing effect, produces conditions which are essentially those of a co-current system. In order to secure a counter-current effect it would be necessary to have more than one fluidised bed in series—either by using separate containers or by segregating or superimposing a series of containers having a fluidised bed in each. Further, due to the thorough intermixing of solid particles within the bed, the drying times of the individual particles as they progress from feed to discharge points are not equal, with the result that there may be underdried particles as well as overdried particles in the discharge. The moisture content of the product will be the average of the mixed particles so that with heat-sensitive material it may be difficult to secure absolute uniformity of low final moisture content figures. For extremely accurate control of product quality it may be necessary to employ multi-stage counter-current fluidised beds.

A simple continuous fluidised bed dryer arrangement is illustrated diagrammatically in Fig. 59 (*a*) whilst Fig. 59 (*b*) and (*c*) illustrate a multi-stage column dryer and a multi-stage sectional dryer of a cascading type. The latter (see also Fig. 65) may sometimes be arranged for solids to pass from section to section under instead of over weirs, the weir partition plates being adjustable in order to vary residence time in each section.

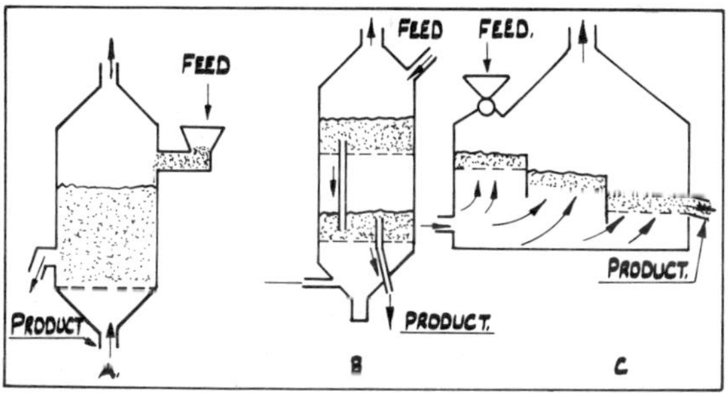

Fig. 59. Continuous fluidised bed dryer arrangements.

The most commonly used design of continuous fluid bed dryer is shown diagrammatically in Fig. 60 as a complete arrangement and a typical industrial installation is shown in Plate XXVII. The dryer is composed of a cylindrical or rectangular section column supported on a gas-mixing chamber extension to an air heater, there being a fluidising plate at the base of the column on which the fluid bed itself is supported. Provision is made in the upper region of the column for introducing a continuous stream of moist charge, whilst a flanged connection at the head of the column leads exhaust gases to a high efficiency cyclone and any further dust recovery equipment found necessary and to an exhaust fan. A high-pressure air blower is generally provided to dilute high-temperature combustion gases from a furnace and with which to fluidise the bed. In some instances small multiclones may be built in integral with the column above the bed and discharging recovered dust into the bed itself. Adequate access holes are provided for emptying and cleaning the interior of the

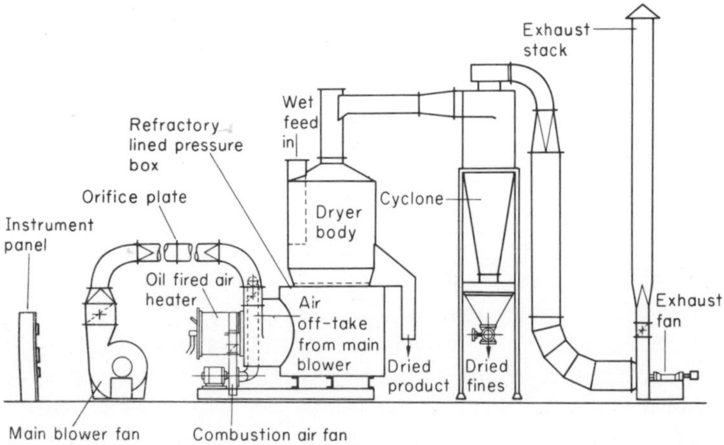

Fig. 60. Typical arrangement of a fluidised bed drying system.

Plate XXVII. Continuous fluid bed dryer installation. *Courtesy Fluostatic Ltd.*

equipment for purpose of maintenance. Dry material is discharged from some point towards the base of the column either over an adjustable weir plate or through a submerged pipe discharge.

When drying is to be carried out at relatively low temperatures the source of heat may be steam or electricity, but due to the extremely rapid fall in temperature of the gases at the distributor plate it is more usual to employ a combustion chamber air heater, generally of similar design to those used with rotary dryers but designed to operate under pressure. Diluent air is introduced into the combustion chamber or into the gas-mixing chamber at the base of the fluidising column or as an extension to the combustion chamber. The temperature of the mixed gases is thermostatically controlled either from the bed temperature or the exhaust gases immediately above the bed. Where the products of combustion may not be brought into contact with the solids, an indirectly-heated heat exchanger is used and, when air cannot be used, a heat exchanger is employed to heat a neutral gas recirculation stream, with provision for extraction of moisture or solvents before return to the heater.

The Distributor Plate

The distributor plate may be of metal or porous ceramic material the nature of which will depend upon the temperature conditions at the plate. During shut-down the plate will have to support the non-fluidised bed and will need to be of a design which will not permit fall-through of material in the bed. These designs may range from bar grids, perforated plate and sintered material to specially designed orifices provided with caps. They will all be designed to give as uniform a distribution of fluidising medium over the area of the distributor plate as possible, to promote uniform fluidisation. The design should also be such as to result in the lowest possible loss of pressure whilst still maintaining uniform distribution. Resistances under operating fluidising gas velocities of the order of 5–6 in w.g. are normal and the relationship between grid pressure loss and pressure loss due to hydrostatic head of the fluidised bed will largely depend upon the operational bed depth and density and may range from 20% on fairly deep beds to 100% for shallow beds. The pressure loss in the distributor plate will be a function of the number and size of orifice and of the fluidising medium velocity. The grid or orifice free area will range over an extremely wide range depending upon the design used and may be from 3–50% of the distribution plate total cross-sectional area. The free area ratio for perforated plates will usually be higher than for capped orifices, a figure of from 5–10% being usual for the latter. For perforated plates considerably thicker than the perforation diameter and for values of 5–10 mm for the latter an empirical relationship between grid pressure drop, fractional free area and operating velocity of the fluidising medium of

$$P = 1{\cdot}44 \times 10^3 \left(\frac{v_f}{A_v}\right)^2$$

has been found[7] where the fractional free area (A_v) falls in the range of 5 to 49%; P is the pressure drop in mm of water and v_f is the operating fluidising gas superficial velocity in m/s.

Capped orifices usually take a form very similar to those employed on fractionating column trays and consist of tubes of short length, either welded or expanded into holes drilled in the distributor plate and provided with conical or domed caps at their upper extremity, beneath which are several drilled holes through which the fluidising medium passes. The free area in this case is given by the expression[7]

$$A_v = 90{\cdot}7 \left(\frac{d_e}{l_k}\right)$$

in which l_k is the pitch of the caps set in a hexagonal arrangement (metres) and d_e is the equivalent hole diameter (metres) defined by

$$d_e = d_k\sqrt{n}$$

where d_k is the hole diameter (metres) and n is the number of holes in one cap. Normal values for A_v will be 2·5–5% of the distributor plate area.

Pressure Loss

The pressure loss in the fluidised bed itself will depend on the density of the solid particles, the density of the fluidised bed which will be a function of the percentage of voids, the bed depth and the velocity of the fluidising gas. The probable pressure loss will have to be determined during pilot-scale tests together with the investigation of other factors for optimum operating conditions. Bed density has a pronounced effect on pressure loss and may range from 60% for a dense phase bed to a nominal loss for dispersed phase fluidisation. However, the latter condition is outside the scope of fluid bed drying technique and is met with in pneumatic dryers. Actual pressure losses in the bed under operating conditions will be those appropriate to a condition between the dense phase and a dilute phase and will be indicated during pilot-scale tests by a reduction in the rate of pressure loss increase as the superficial fluidising medium velocity is increased.

A typical pressure drop curve for a bed of sand is given in Fig. 61 where pressure drop in in. w.g. is plotted against air flow through the bed in

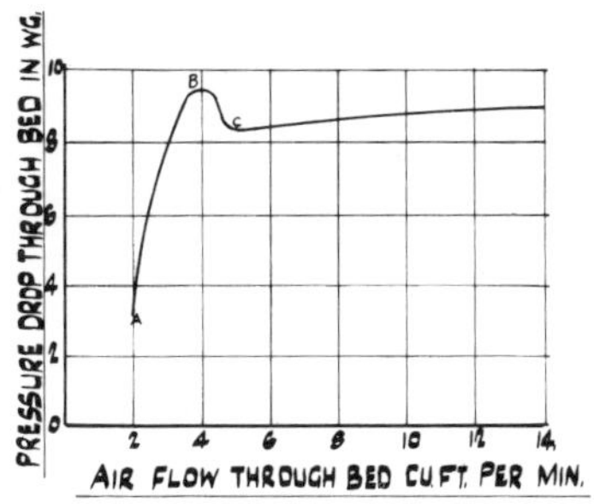

Fig. 61. Typical pressure-drop curve for fluidised bed of sand.

ft^3/min. From point A to BC the air is passing through the bed of sand without causing movement of individual particles. Increase of air-flow beyond B causes expansion of the bed, considerable increase in voids and a consequent reduction of pressure loss between points B and C. Thereafter the pressure drop increases only slightly with increase in air-flow rates, complete fluidisation having been attained. The operational superficial velocity will be that at which any entrainment of material is such as can be tolerated, unless the dryer is deliberately intended to act as both a dryer and a classifier. Where high initial gas temperatures are not permissible it may be necessary to operate at rather higher superficial gas velocities than those required for promoting optimum fluidisation conditions, in order to supply the necessary thermal energy for the drying operation. Carry-over by entrainment would not normally be acceptable above 5–10% of the material feed rate (dry basis) unless the sieve analysis of the feed covered a wide range, in which case it might be as high as 20%. The carry-over in any event will consist of the smallest particles which dry more easily and these can be blended in with the main dry product discharge by feed-back from the dust recovery system. Larger particles, even if carried out of the bed by the fluidising gas, generally fall back due to the loss of gas velocity in the free space above the bed. The fluidised bed depth should not be greater than the bed diameter and it is usual to allow not less than 3–6 ft of clear height above the fluidised bed before the exhaust gas exit point.

It is found that for feedstocks composed of relatively large particles the range of practical operating velocities between incipient fluidisation and entrainment is narrower than for feedstocks composed substantially entirely of fine particles; the ratio of incipient fluidising velocity and entrainment velocity decreases with particle size. The degree of entrainment may also be influenced by the initial and final moisture contents due to the longer requisite residence time in the bed to attain complete drying from high initial moisture feedstocks, with consequent entrainment of the finest particles, which are dried more quickly, are lighter and may suffer from a measure of attrition.

A useful chart for estimating superficial fluidisation velocities is given by Clark[8] based on a rule of thumb developed by W. C. Lapple[9], which states that the operating velocity is one-tenth the terminal settling velocity (free-fall velocity) for particles up to 10 mesh and one-fifth the terminal settling velocity for larger particles. This chart is reproduced in Fig. 62, noting that particle diameter is represented by weight average size as determined by sieve analysis. Operating superficial gas velocities usually lie in the range of two to three times the superficial incipient fluidising velocity.

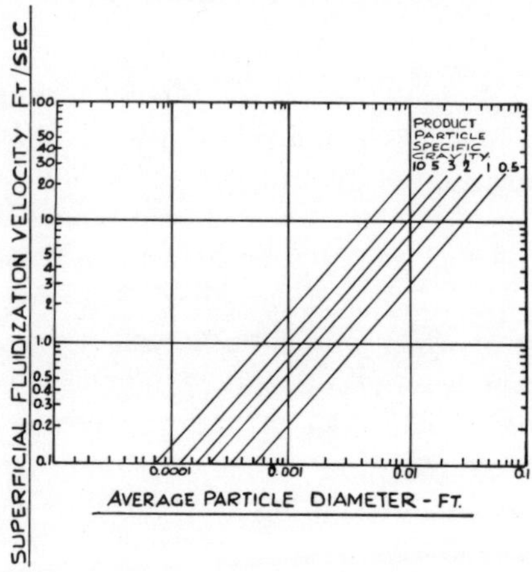

Fig. 62. Fluidisation velocity—particle size—density relationship chart.

(after Clark, E. W. Chem. Eng. 13 Mar. 1967. P. 180)

Continuous Fluid Bed Dryer Design

The selection of the most suitable superficial fluidising medium velocity will also be influenced by the required evaporative capacity of the dryer in relation to the grid area, i.e. its specific drying capacity. This increases with operating velocities due to higher thermal inputs per unit of time and higher rates of mass and heat transfer and will have its effect on the necessary residence time for the drying operation which, in turn, will be affected by the temperature of the entering gas.

Pilot-scale tests will determine the permissible operating temperature, acceptable gas velocity and the residence time for an optimum combination for the first-mentioned two variables. As the efficiency of the dryer will depend on the temperature difference between the ingoing gas and that of the bed (which is very approximately that of the exit gas), the highest possible inlet gas temperature will be selected compatible with acceptable bed temperature and the cost of materials of construction for the grid and other equipment at the hot gas feed point. However, as at large differences between gas inlet and bed temperatures increase in gas inlet temperature has a decreasing effect on specific heat consumption by the bed, an upper limit of 1,500–2,000°F is normal for non-heat-sensitive materials. Most direct-fired fluid bed dryers of mild steel construction have usually a maximum temperature range for inlet gases of 900–1,400°F.

As the residence time of material in the dryer is the quantity in the bed divided by the production rate, and as good practice indicates that the fluidised bed depth should not be greater than the bed diameter, an essential factor in designing a fluidised bed dryer is the calculation of a minimum bed diameter. This involves the relationship between the diameter of the bed, the operating velocity and the exit gas relative humidity. As the exit gas will leave the bed at substantially the same temperature as that of the bed, the relative humidity of the gas leaving the bed should not be higher than that corresponding to the equilibrium moisture content of the driest particles at that temperature. It is usual therefore in designing fluid bed dryers to determine the equilibrium moisture curve for the product at the bed temperature selected from pilot-scale tests and from this curve to ascertain the relative humidity of exit air corresponding to the desired terminal moisture content of the product. The bed diameter should be only slightly larger than that necessary at the chosen air velocity to provide exit gas having a relative humidity at bed temperature equal to the humidity figure arrived at from the curve. The chosen air velocity will usually be about twice that required for incipient fluidisation. A material balance will determine the moisture content in the entering and exit gases, noting that for direct-fired air heaters the moisture produced by the combustion of the fuel must be included. The material balance equation will be:

$$\frac{(H_r)}{100}\frac{(p_s)}{p_c} = \frac{W+[M/(1+H_e)]H_e}{W+[M/(1+H_e)]\,(0{\cdot}625+H_e)}$$

where H_r = relative humidity of gases leaving the bed
p_s = vapour pressure of water at gas exit temperature, atmos.
p_c = total static pressure of gas leaving bed, atmos.
W = moisture removed in drying, lb/h
M = flow rate of fluidising gas entering bed, lb/h
H_e = humidity of fluidising gas entering bed, lb/lb dry gas
0·625 = ratio of molecular weight of water to that of dry gas.

For normal conditions, the value of p_c will be unity, whilst the expression $M/(1+H_e)$ is the flow rate of bone-dry air lb/h. The value of p_s is given in vapour pressure tables or from the chart in Fig. 63, it being normally assumed that the exit gas temperature is approximately that of the fluidised

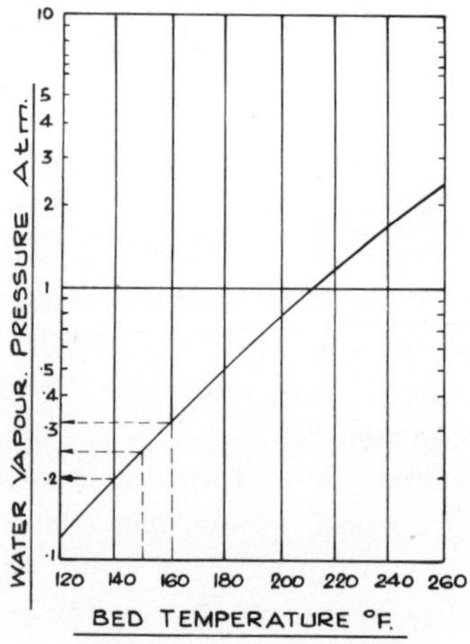

Fig. 63. Water vapour pressure—bed temperature relationship.

bed as determined in pilot-scale tests for a given inlet gas temperature and product feed rate.

A heat balance on the system will take the form:

$$0{\cdot}25M(t_m - t_b) = \lambda W + c_f F(t_b - t_f)$$

Where 0·25 = specific heat of fluidising medium approximately, Btu/lb/deg F (0·24 for air in indirect air heaters)

t_m = temperature of fluidising gas entering bed, °F

t_b = temperature of fluidised bed, °F

t_f = temperature of wet feedstock, °F

λ = latent heat of evaporation of water at t_b, Btu/lb

c_f = average specific heat of wet feedstock, Btu/lb/deg F

M = flow rate of fluidising gas entering bed, lb/h

F = flow rate of wet feed stock, lb/h

W = moisture removed in drying, lb/h

The value of c_f will be calculated from the moisture and solids content of the feedstock and their respective specific heats. Simultaneous solution of these equations will give the value of M and t_b once an inlet air temperature has been selected from pilot-scale tests. Having determined M, Clark[10] derives the minimum bed diameter D:

$$D^2 = \frac{(M + 1{\cdot}58W)(t_b + 460)}{111{,}000 v_f}$$

where D is the bed diameter in feet and v_f is the fluidising velocity in ft/s obtained by experiment. The reader is referred to Clark's paper and to *Fluidised Bed Drying* by Vanecek, Markvart and Drbohlav (Leonard Hill

Books, 1966) for a more detailed and informative treatment of continuous fluid bed design.

The steps in designing the continuous fluid bed dryer will therefore be:

1. Establish that the feedstock can be fluidised.
2. Determine by pilot tests the optimum fluidising velocity, inlet air temperature and residence time for adequate drying.
3. Determine the drying rate curve and the equilibrium moisture content of the product at the probable bed temperature indicated by (2).
4. Estimate the probable exit gas relative humidity from (3).
5. Calculate the bed temperature and required fluidising medium flow rate by simultaneous resolving of the material and heat balances appropriate for the method of heating.
6. Calculate the minimum bed diameter.

The test for establishing whether the material can be fluidised and its behaviour at different fluidising medium velocities, including an indication of the bed depth-diameter relationship during fluidisation, is usually carried out in a glass tube of 3–4 in diameter fed with a supply of air. For actual pilot-scale tests using a continuous feed of wet material the fluidising column, generally of mild steel and provided with observation windows, should be not less than 12 in in diameter and 5 or 6 ft high above the distribution plate, which should be substantially of the same design that is intended for the production plant. Using a gas velocity which gives satisfactory fluidisation and a bed depth of about 12 in, a series of tests will establish the optimum relationship between material feed rate and inlet gas temperature compatible with acceptable bed temperature and final moisture content of the product. The result gives the evaporational capacity per square foot of bed cross-sectional area. Further tests at different bed heights and fluidising velocities will establish the most economic combination, so that evaporational capacity does not increase materially with increased bed depth. This bed depth will determine the minimum bed diameter, and the corresponding evaporational capacity per square foot of bed cross-section can be used for scale-up purposes. The height of clear space above the depth of fluidised bed should be not less than about 1·5 times the latter to permit of disentrainment of solids.

For most drying operations within industry the normal method of supplying the necessary thermal input is by a direct-fired air heater of the combustion chamber type using either liquid or gaseous fuel. Due to the pressure losses caused by the distributor plate, the bed itself and the solids-gas cyclone separator, the heater is usually designed to operate under pressure. When the material is exceptionally heat-sensitive an indirect air heater will be used. On the other hand, if the product is extremely resistant to high temperatures and where a high bed temperature is desirable, the fuel may be fed to a number of points at the base of the bed, combustion taking place in the bed itself, fluidising and combustion air – usually pre-heated by the outgoing gases in the upper free space of the dryer column – being introduced in the usual manner below the distributor plate. The bed itself must, of course, be above the ignition point of the fuel initially, and as the ignition point of liquid fuels is lower than natural gas they are usually preferred. A state of surface combustion is set up in the bed and, due to the high degree of turbulence which exists, rapid and complete combustion takes place with immediate transfer of heat to the charge. A dryer or dryer-calciner operated on this system would usually pass the hot dry product to a second stage fluidised column acting as a cooler, using the fluidising and combustion air for the high temperature stage as coolant. For high-temperature operations the distributor plate would be constructed in heat-resistant ceramic material or alloy and the column itself would be refractory-lined. External thermal insulation of the dryer itself, interconnecting ductwork and the cyclone is important in order to reduce heat losses to a minimum and to prevent condensation in the cyclone.

Operation and Control

The operation of continuous fluidised bed dryers, although relatively simple, does call for careful supervision. In starting up, the direct-fired air heater is usually brought up to a sufficiently high temperature to attain correct combustion conditions, the products being exhausted through an auxiliary stack. The main air blower is then brought into service, the mixed air and combustion product being led via the mixing chamber through the distributor plate to the drying column itself and the desired inlet air temperature attained. Wet feedstock is now introduced continuously to the column and dried product is removed when the desired bed depth has been attained. Feed rate, product discharge rate, fluidising medium quantity and temperature are then progressively increased until steady-state conditions at the required feed rate, inlet temperature and product moisture content are established. This method of start-up is applicable to free-flowing materials but for others it is usual to feed pre-dried material at a moderate rate to the column with hot fluidising medium flowing at a moderate rate until the required bed depth has been established. The fluidising medium feed rate is then raised to the desired figure and wet feedstock is introduced when the fluidised bed has attained the desired temperature. Feedstock rate and fluidising medium temperature are then raised to the required extent and dried product is removed at the necessary rate to maintain a constant fluid bed depth. Adjustment in feed rate and fluidising medium temperature may have to be made so that the temperature of the exhaust gases leaving the cyclone is above dew-point. During the operation of the dryer, once conditions have stabilised it is important to maintain material feed, discharge rate and fluid bed temperature constant at the desired values.

In reverse, for shutting down the dryer, the wet feedstock and source of heat are shut off, the supply of fluidising medium being meanwhile continued. Once the bed temperature has fallen adequately it can be emptied and the fluidising medium fan then stopped. If the design of distributor plate is such that any remaining material in the column falls through it on shutting down the fan, this is removed from the chamber below the column before any further start-up.

Uniform feed rate of moist material and its close control are essential for good operation and the feeding device should be chosen accordingly. Further the design of feeder should be such as to create a reasonable seal unless the atmosphere above the bed is balanced with the outside atmosphere.

Control of operation of a continuous fluidised bed dryer is almost invariably by automatic methods because of the difficulty in controlling conditions manually in view of the high rate of drying and the low thermal lag of the fluidised bed. The most satisfactory method is to maintain the fluidising medium temperature and quantity constant; maintain the fluidised bed temperature constant by controlling the feed rate; maintain the fluid bed depth constant by controlling the dry product discharge rate. Variation in feed rate for a constant bed depth will of course affect the residence time, but since the most usual cause of a fall in bed temperature is an increased moisture content in the feed, such a drop in bed temperature will decrease the feed rate of wet material and increase the residence time. Conversely, a fall in initial moisture content will lead to an increase in bed temperature which in turn, under automatic control, will increase the feed rate and decrease the residence time. The fluidising medium quantity being set at a constant rate, any variation in the material feed rate will affect the resistance (hydrodynamic) of the bed which in turn will tend to affect the flow of fluidising medium. The resistance of the bed must be kept constant by controlling the rate of discharge of dry product. The following automated controls will therefore usually be used in operating a continuous fluid bed dryer:

1. Fuel consumption or steam supply: thermostatic control.
2. Fluidising and combustion air: orifice and pressure control.
3. Feedstock: motorised thermostat from fluid bed.
4. Fluidising medium and fluid bed temperature: thermostatic control.
5. Dry product discharge: manometric from bed pressure loss.

Continuous or intermittent recording of the following variables should also be provided:

1. Fuel or steam consumption.
2. Fluidising medium.
3. Material: feed rate; dry product; cyclone recovery.
4. Temperatures of
   (a) drying medium below grid,
   (b) combustion gas before admixture with excess air for fluidising,
   (c) fluidised bed,
   (d) drying medium leaving dryer,
   (e) drying medium leaving cyclone.
5. Pressure drop
   (a) across distributor plate and bed,
   (b) across cyclone and any other dust-recovery equipment.
6. Moisture contents of
   (a) feedstock,
   (b) dry product.

The desirable measurement and control points for a fluidised bed drying system are shown diagrammatically in Fig. 64.

Typical performance data for continuous fluidised bed dryers as shown in Fig. 60 are given in Tables 30–32 for a variety of products, and in Table 33 for dryers of the type shown in Fig. 65.

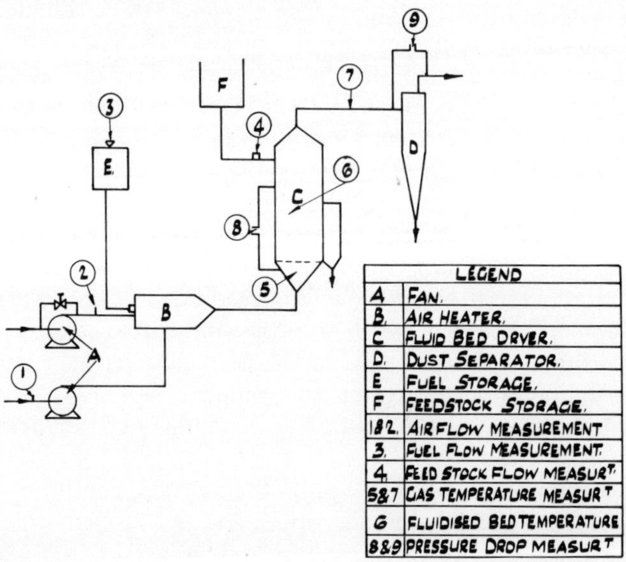

Fig. 64. Variable measurement and control points on a continuous fluid bed drying plant.

(after Vanecek, Markvart & Drbohlar)

Table 30. Performance data on continuous fluid bed dryers

	Coal	*Silicious grit*	*Glass sand*	*Sand*	*Asphalt*
Method of feed	Screw Feeder	Chute	Chute	Chute	Chute
Material size	$-\frac{1}{2}$ in	$-\frac{1}{16}$ in	−36 mesh	$-\frac{1}{16}$ in	$-\frac{3}{16}$ in
Product rate, lb product/h	190,000	17,920	15,680	33,600	22,400
Initial moisture, % w/w basis	14	5	7	5	5
Final moisture, % w/w basis	7	0	0	0	0·5
Residence time, min	2	$1\frac{1}{2}$	3	3	10
Dryer diameter	7 ft 3 in	3 ft 0 in	4 ft 6 in	6 ft 6 in	8 ft 0 in
Fluid bed height, in	21	12	12	12	24
Air inlet temperature, °F	1,000	1,400	1,400	1,400	470
Air outlet temperature, °F	135	230	230	230	220
Air quantity, ft^3/min Std	20,000	2,000	2,000	3,500	7,000
Material exit temp, °F	140	230	230	230	220
Evaporation rate, lb/h	11,200	896	1,097	1,680	1,120
Method of heating	Coke-oven gas	Gas oil	Town Gas	Gas oil	Gas oil
Heat consumption, Btu/lb water evaporated	2,000	2,250	2,000	2,200	1,800
Fan installed, hp	210	$32\frac{1}{2}$	18	30	90

Courtesy Head Wrightson Stockton Ltd.

Table 31. Performance data on continuous fluid bed dryers

	Abrasive grit	*Clay granules*	*Sand*	*Granular desiccant*	*Household salt*
Product rate, lb/h	2,200	1,000	14,000	150	13,500
Initial moisture, % w/w basis	9	22	6	25	4
Final moisture, % w/w basis	dry	3	dry	7	0·03
Air inlet temperature, °F	580	160	325	300	390
Air outlet temperature, °F	210	120	140	205	230
Method of heating	Gas	Steam	Gas	Gas	Steam
Heat consumption, Btu/lb water evaporated	2,700	3,800	2,700	3,600	5,100
Bulk density, lb/ft^3	120	60	90	30	60
Average drying time, min	2·5	30	3	24	4
Fan capacity, ft^3/in^2	2·5	1·35	1·05	0·84	1·05
Installed fan hp	10	45	25	5	50

Courtesy Pennsalt Ltd.

Table 32. Performance data on continuous fluid bed dryers

	Coal	*Sand*	*Silica sand*	*Limestone*	*Iron ore*
Material size	$\frac{1}{2}$ to 0	−25 to 0	−18 to 0	$\frac{3}{16}$ to 0	$-\frac{3}{8}$ to 0
Method of feed	Twin screw	Bucket elev.	Conv.	Conv.	Conv.
Product rate, lb product/h	448,000	22,400	112,000	67,000	896,000
Initial moisture, % w/w basis	11	6	6	15	3
Final moisture, % w/w basis	5·5	0·1	0·1	0·1	0·75
Residence time, min	1	1·25	1·5	1·25	0·5
Dryer diameter, ft	10	3·0	7·25	5·5	8·5
Fluid bed height, in	18	12	12	12	18
Air inlet temperature, °F	1,000	1,200	1,200	1,200	1,200
Air outlet temperature, °F	170	212	212	212	212
Air quantity, ft^3/min Std	40,000	2,000	9,000	13,000	45,000
Material exit temperature, °F	140	220	220	220	220
Evaporation, lb/h	24,640	1,430	6,720	11,880	20,400
Method of heating	Coal	Gas	Oil	Oil	Oil
Heat consumption, Btu/lb water evaporated	1,830	1,620	1,730	1,220	2,300
Fan installed hp	240	20	80	115	350

Courtesy Fluostatic Ltd.

Table 33. Performance data on continuous fluid bed dryers

	Sodium perborate	*Weed killer*	*P.V.C.*	*Coal*	*Sand*
Method of feed	Screw	Vibrator	Screw	Vibrator	Vibrator
Material size	30–200 mesh	5–1 mm flake	60–120 mesh	3 mesh–zero	30–120 mesh
Product rate, lb product/h	11,400	5,100	10,075	440,000	112,000
Initial moisture, % w/w basis	3·5	14	2·0	8	8
Final moisture, % w/w basis	0·0	0·2	0·2	1	0·2
Residence time, min	1·5	11	30	0·3	0·45
Drier bed size, ft × ft	22·5 × 5·5	18 × 4·5	23 × 6	16 × 6·6	12·5 × 3·2
Fluid bed height, in	4	3	18	5	6
Air inlet temperature, °F	176	212	167	932	1,202
Air outlet temperature, °F	104	150	122	180	221
Air quantity, ft^3/min Std	6,600	14,200	5,400	67,330	8,000
Material exit temperature, °F	104	205	122	180	212
Evaporation, lb/h	400	720	183	33,440	9,750
Method of heating	Steam	Steam	Steam	Coke-oven gas	Oil
Heat consumption, Btu/lb water evaporated	2,100	3,060	4,640	1,970	2,200
Fan installed hp	33	40	34	600	70

Courtesy Rosin Engineering Ltd.

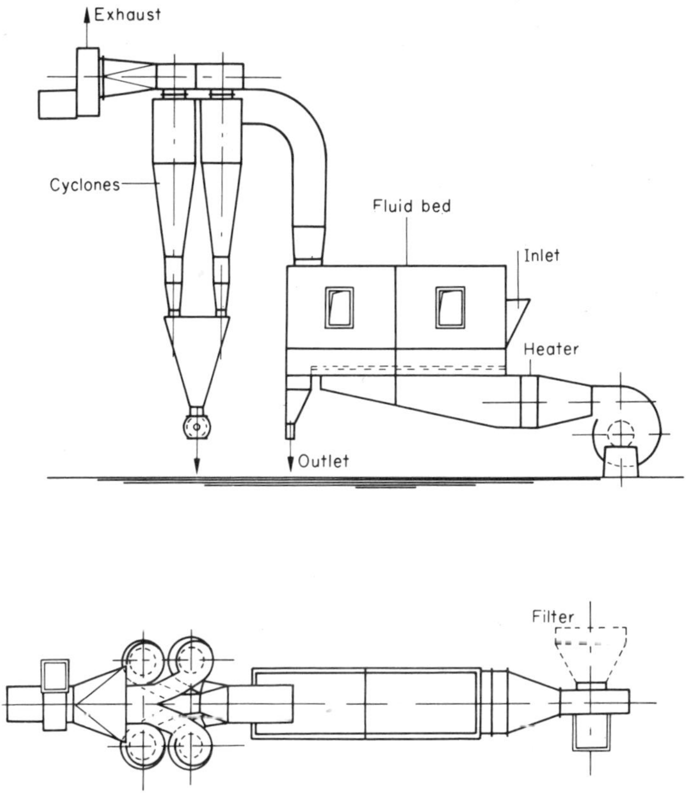

Fig. 65. Buttner–Rosin multi-stage sectional continuous fluid bed dryer.

As with any type of equipment, standardisation of design and fabrication methods leads to reduction in capital costs where the demand justifies repetition manufacture. The dryer illustrated in Fig. 66 is supplied in three standard sizes all of the same design having a built-in cyclone discharging to a point in the fluidised bed adjacent to the main dry product discharge. This design results in considerable conservation of factory floor space and structural steel support work and the three sizes of dryer cover a production range of five to forty short tons per hour depending upon the evaporational load. Fig. 67 gives a series of three curves indicating the normal performance of each size of dryer for a specified moisture removal at normal combustion gas inlet temperature.

Dryer size number	Dimensions						Motor hp		
	A	B	C	D	E	F	Main fan	Exhaust fan	Compressor
1	17'-0"	11'-0"	3'-9"	7'-6"	9'-0"	3'-3"	20	10	1
	5181·8	3352·8	1143·0	2286·0	2743·2	990·6			
2	22'-3"	15'-0"	7'-7"	7'-6"	12'-0"	3'-9"	40	20	3
	6781·8	4572·0	2311·4	2286·0	3657·6	1143·0			
3	26'-0"	18'-0"	8'-7"	9'-6"	14'-0"	3'-9"	65	45	5
	7924·8	5486·4	2616·2	2895·6	4267·2	1143·0			

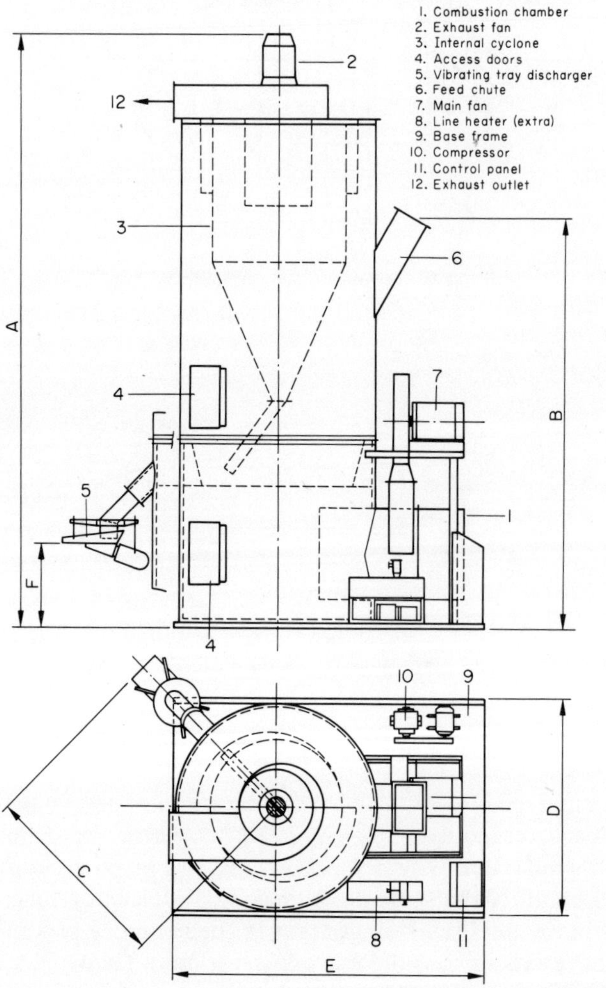

Fig. 66. "Fluostatic" standard continuous fluid bed dryer. *Courtesy Fluostatic Ltd.*

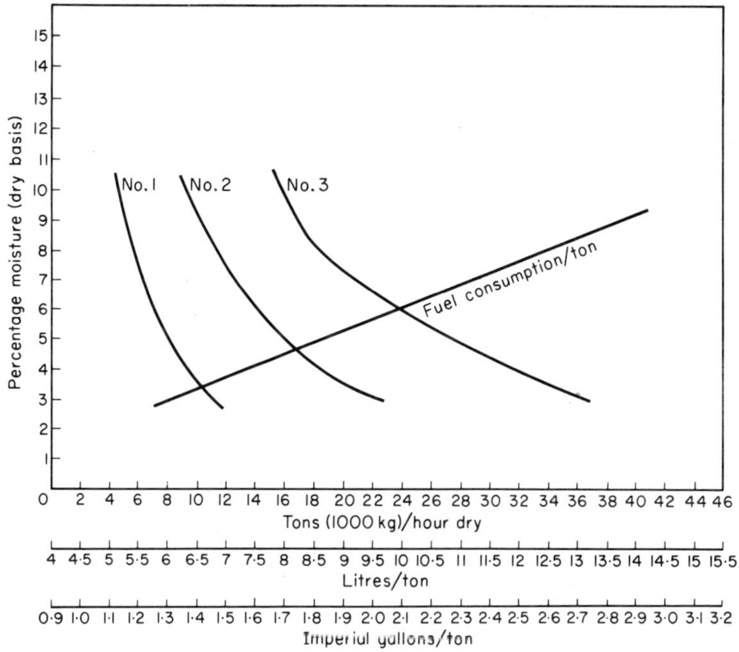

Fig. 67. Performance curves for "Fluostatic" standard range continuous fluid bed dryer.

Courtesy Fluostatic Ltd.

CHAPTER X

Spray Dryers

Solutions, suspensions, slurries and pastes which are pumpable are normally dried on either film-drum dryers or in spray dryers. Whereas the drying operation on a film drum dryer involves contact between the feedstock and the heated drum for an appreciable length of time to complete drying depending upon the speed of revolution of the drum – usually 5–12 rev/min – evaporation in a spray dryer is almost instantaneous, the drying medium temperature undergoes rapid reduction and the dried material is not raised above this terminal medium temperature. Spray dryers are therefore ideally suitable for materials which are in any way heat-sensitive. Much higher drying medium initial temperatures are permissible than with film-drum dryers, considerable temperature reductions occur and, in consequence, evaporative capacities and thermal efficiencies can be high. Contact times between feedstock and drying medium are very short, the atomised liquid drops are kept cool by the high rate of evaporation, whereas when evaporation has terminated the dried particles are in contact with considerably cooled gases, thus eliminating danger of thermal degradation of the product. With film-drum dryers, on the other hand, the dried product remains substantially at the temperature of the drum surface until removed by the doctor knives.

The physical nature of the dried product delivered by a spray dryer differs from that of a film-drum dryer and may be a more desirable marketable product. Film-drum dryers usually yield the product in the form of flakes, film strips or as a fine powder, whereas the product from a spray dryer will usually be in a sphere-like form the dimensions of which are amenable to a certain degree of control. The flow characteristics, relative freedom from dust and suitability for packaging of the latter may be important properties in its favour. The spherical particles of dried product may be solid or hollow, depending on the drying conditions and nature of the feed. Solutions from some materials form droplets which on drying produce a case-hardened outer surface and, as a result of the rapid evaporation of the internal moisture, expand to form a hollow sphere which may become punctured by the released moisture. This phenomenon will often have a bearing on the bulk density of the dried product and it is usual to investigate the correct relationship between solids content of the feedstock, method of atomising, inlet gas temperature and relative direction of flow of product and drying medium to attain the most desirable dry product bulk density. Particle size will usually depend on type and operation of atomiser device and spray chamber configuration as well as relative movement of product and drying medium. From the point of view of packaging, bulk density is of major importance, and in general it may be said that bulk density can be increased by increased solids content of the feed, lower inlet gas temperatures, counter-current mass flow, fine atomisation and by the use of certain additives in the feedstock.

Spray drying has many advantages over film-drum drying apart from the question of short contact times and reasonable dry product temperatures. Spray dryers can be designed for small productions up to very considerable evaporative capacities – 12,000–15,000 lb/h not being unusual – whereas film dryers are limited by the physical size of a unit which can be manufactured. A 300 ft^2 surface area twin-drum dryer would be considered a large unit and might have an evaporative capacity under favourable conditions of say 3,000 lb/h. Maintenance costs on spray dryers are low as there are few moving parts and corrosion is reduced to a minimum because the wet material is usually only in contact with the atomising device, which can be manufactured in suitable resistant materials. Purity of the product will be high as there is little or no danger of pick-up of foreign substances – which may occur with film drum dryers through the action of the doctor knives on the heated drum unless knives of special materials are practicable. Labour costs are low due to the high degree of automation possible; one operator can supervise even the largest installation. It is more easily possible to apply an installed spray dryer to a range of products than a film dryer, due to the difficulty of cleaning the latter from over-dried product or wet feedstock solids in the feed system. Spray dryers can be started up and shut down quickly and steady-state operating conditions are rapidly attained. Whereas it would be virtually impracticable to apply a film dryer to the evaporation of a volatile or inflammable solvent or vehicle unless it were designed as though for vacuum operation, a closed-cycle spray dryer does not raise any serious problem. Dust problems, so often present with film dryers, do not arise with spray dryers which have built-in dust recovery equipment.

A typical spray dryer flow sheet is shown diagrammatically in Fig. 68

Plate XXVIII. Industrial spray drying installation. *Courtesy Bowen Engineering Inc.*

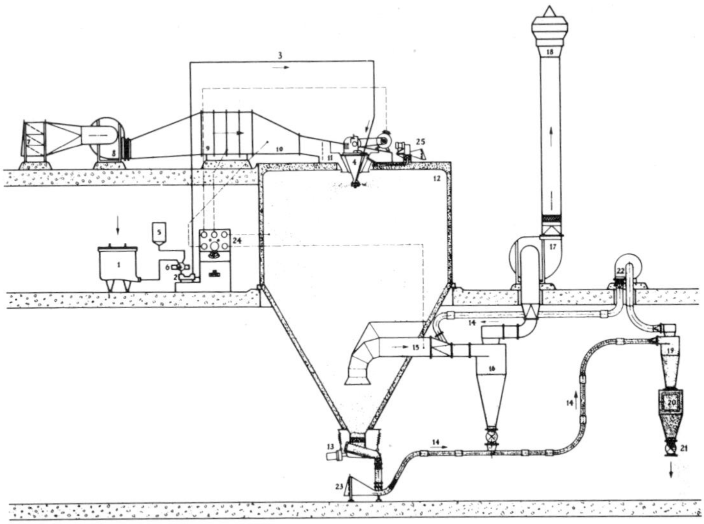

Fig. 68. Typical flow diagram of spray dryer. *Courtesy Niro Atomizers Ltd.*

and a typical commercial installation in Plate XXVIII. Essentially a spray-drying plant consists of a source of hot gases, a drying chamber, means of atomising a pumped source of feedstock, provision for withdrawing the dried product and exhaust gases from the drying chamber and equipment for the separation of the dried product from the exhaust gases.

The source of hot gases may be one of the many used on convection dryers depending on the permissible maximum inlet gas temperature, whether the product may be brought into contact with the products of combustion or with air and on the freedom from contamination of the product. Low inlet gas temperatures will usually involve steam or fuel fired indirect heaters, as will the use of neutral drying media on closed cycle operation or where volatile and inflammable components are present. For most operations direct-fired combustion chamber air heaters are used, natural gas, manufactured gas and oil being the most usual fuels, particular attention being paid to the furnace design to promote complete combustion. Inlet gas temperatures will usually range from 200°–1,400°F depending on the heating method employed. Due to the very high evaporative capacity of large spray-dryer installations, the use of high intensity combustion air heaters is often desirable in order to restrict the size and capital cost of the heating equipment.

The extremely high evaporation rates obtainable in spray dryers are due to the high ratio of surface area to mass of the droplets produced in the atomising device. Further, relatively uniform particles are desirable because over-sized droplets will tend to approach the walls of the drying chamber before adequate drying has been secured, resulting in build-up on the walls. Conversely, under-sized droplets will cause fine dust which may become overheated. The design and operation of the atomising equipment is in consequence of major importance. Three methods of atomisation are normally employed, namely, pressure (or single-fluid) atomisers; pneumatic (or two-fluid) atomisers; and centrifugal atomisers.

Pressure atomisers, Plate XXIX, use energy provided by high pressure pumping equipment delivering the feedstock to an atomising nozzle having a small orifice, the feedstock being given a high degree of spin by its action. Pressures employed may range from 100–10,000 lb/in^2 although the more normal range is from 300–4,000 lb/in^2. Orifice sizes will range from 0·012–0·15 in. The higher the pressure for a given orifice size the greater the atomisation. Pressure atomisers are not suitable for abrasive slurries because of excessive wear set up in both atomiser and pumping equipment

Plate XXIX. Spray dryer pressure atomiser. *Courtesy Anhydro A/S*

under operating conditions. There is also the danger of the orifice becoming blocked. On larger installations multiple nozzle arrangements are used both to cope with the large volume of feedstock and to give uniform distribution of the spray within the drying chamber. Wear by erosion is always possible and, in consequence, orifices are made of hard alloys and even synthetic sapphires. Maintenance on pressure nozzles is usually high as they have to be in near-perfect condition to secure proper atomisation. Product characteristics are affected both by orifice size and pressures employed. Pressure will also affect the capacity of the dryer. Reduction in pressure will coarsen the spray so that in the event of requiring a reduction in throughput, the orifice size should be reduced if the pressure is lowered.

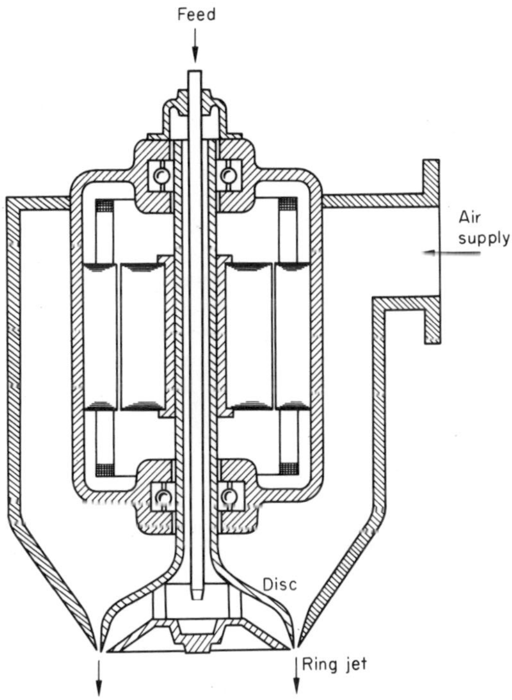

Fig. 69. Pneumatic atomiser. *Courtesy Pennsalt Ltd.*

A pneumatic atomiser, Fig. 69, is normally used only on small capacity and pilot-scale dryers as it does not operate very efficiently at high capacities. It is, however, effective in dealing with thick slurries and pastes and, when manufactured in special abrasion-resistant materials, is applied widely in drying ceramic suspensions. It operates on very much the same principle as paint-spray guns, there being two alternative arrangements, namely, outside and inside mixers, depending on whether the secondary fluid (usually air or steam) is applied around or within the feedstock stream. Feedstock pressures usually range from nil to 60 lb/in^2, whilst the secondary fluid pressure will range from 60–100 lb/in^2, at which pressures atomisation to 10–20 micron size is usually feasible.

A centrifugal atomiser, Fig. 70, uses the energy of centrifugal force set up by a spinning disc or paddle when the feedstock is pumped or fed by gravity thereto. The atomising disc or wheel is designed in such a way as to bring the liquid feed out to the periphery and to attain the speed of the

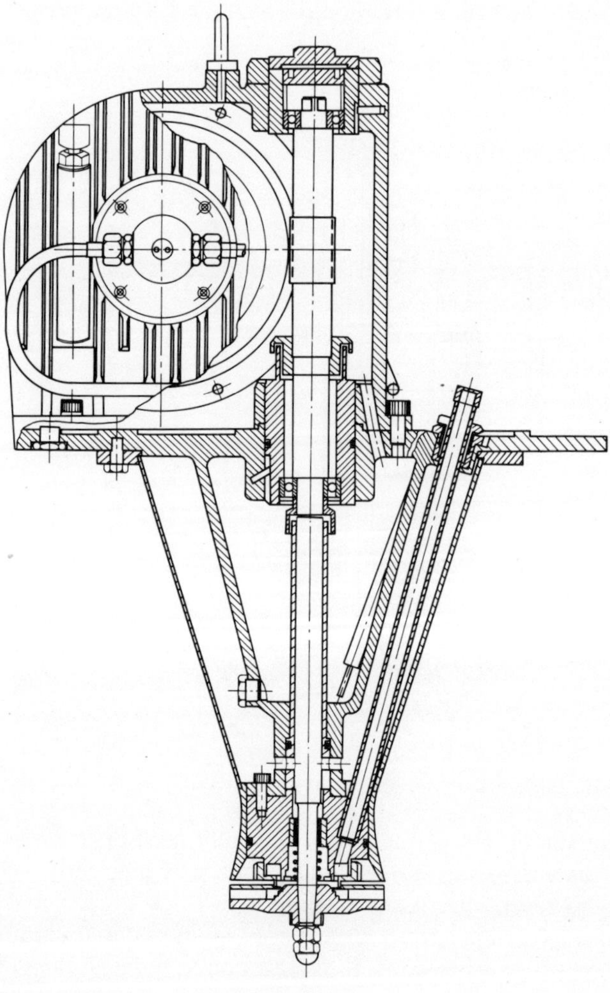

Fig. 70. Centrifugal atomiser. *Courtesy Anhydro A/S*

periphery. The liquid flow becomes accelerated and ejected from the rim at very high speed – usually from 200–600 ft/s – whereupon the liquid film breaks down to droplets. Individual disc and wheel designs have as their objective the formation of as narrow a range as possible of droplet size and attainment of suitable peripheral velocities for the particular feedstock handled. A narrower range and greater uniformity of droplet size at high feedstock input rates is usually obtained with a centrifugal atomiser than with two fluid nozzles – capacities as high as 50,000–60,000 lb/h can be reached with a single large centrifugal atomiser. Disc diameters may range from 2 in for small units to 14–15 in on commercial plants although larger diameters are used in very high capacity installations. Disc speeds may range from 3,000–50,000 rev/min although the more usual plant speeds lie between 4,000 and 15,000. Herring and Marshall[(11)] found that at peripheral speeds between 109 and 305 ft/s, using an aqueous feed, the droplet size varied inversely as the peripheral speed to the power 0·83; at higher speeds (325–496 ft/s) Masters and Mohtadi[(12)] found that the droplet size varied inversely as the 0·7 power of the peripheral speed and directly as the 0·18 power of the feed rate which compared favourably with results at lower speeds found by Friedman, Gluckert and Marshall.[(13)] Centrifugal atomisers are not subject to erosion and blocking to the same extent as nozzles so that they are more suitable for slurries and pastes than the latter. They are more flexible as regards variations in feed rates than nozzles which should be operated under constant conditions to give a required degree of atomisation at specified capacities. Centrifugal discs may be belt-driven, direct-driven by high-speed electric motor powered by a frequency converter or may be steam-turbine driven. The direct-driven electric motor with frequency converter has the advantage of attaining very high disc speeds which can be closely controlled to yield flexibility as regards feed rates and droplet size (and hence product particle size).

Centrifugal discs may vary in design from one manufacturer to another, but the three most usual types are illustrated on Plate XXX which shows a smooth saucer-shaped disc, a vaned disc and a slotted disc. The saucer design may have a rounded or sharp circumferential edge depending on whether a low or high shear action is required and hence coarser or finer atomisation. It is of relatively low cost and is easily replaced when this becomes necessary. The vaned disc, built in a similar manner to the runner in a centrifugal fan, the vanes of which may be forward or backward curved, relies on the outward flow of liquid over the edges of the vanes and is capable of fine atomisation. Maximum degree of atomisation is obtained with the slotted design of disc, when a very fine product is desired or when a highly viscous liquid feedstock is handled requiring a high degree of atomisation. Several variants on the above three main types are manu-

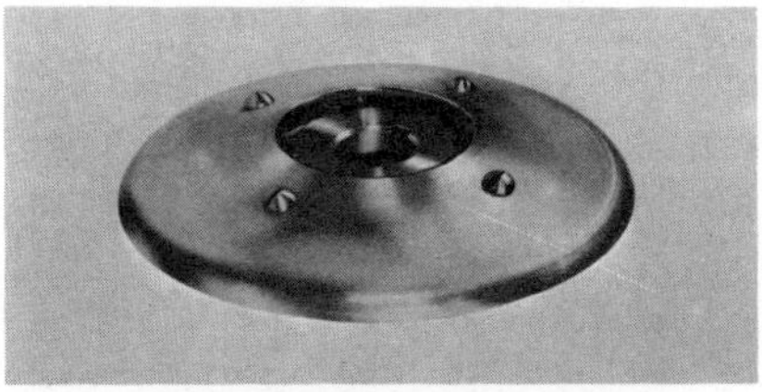

Plate XXX. Centrifugal disc atomisers.
Courtesy Bowen Engineering Inc.

factured and may be supplied in multiple-tier form for large capacities. The choice of design is usually determined by pilot-scale work on the proposed feedstock and a study of the nature of the product produced under test conditions.

Drying chamber design, second in importance to the design, selection and operation of the atomising equipment, will depend upon considerations of direction and degree of atomisation, air-flow pattern, desired product characteristics, retention time for drying, air-flow rates and product discharge requirements. The basic design will be determined by the direction which the atomised droplets will take on ejection from the atomiser; a narrow cone of spray from a nozzle will call for a tall tower chamber to give adequate vertical distance; a horizontally ejected spray from a centrifugal atomiser may require a chamber of large diameter and relatively low height. Air-flow patterns may be co-current, counter-current or a combination of both and may be either straight line or spiral flow. Typical chamber shapes for nozzle atomisation are shown in Fig. 71, whilst those employing centrifugal discs are shown in Fig. 72.

The arrangement shown in Fig. 71(*a*) has a co-current air flow and is widely used for highly heat-sensitive products such as fruit purees, coffee, tea, etc. Due to the height of chambers of tower design, retention times can be maintained adequate for materials with slow-drying characteristics and the formation of a product of large particle size is possible. The counter-flow air pattern, Fig. 71(*b*), is widely used for detergents but in general is applied to materials of relatively high density as the particles have to fall through an upward flow of air. Fig. 71(*c*) illustrates diagrammatically a

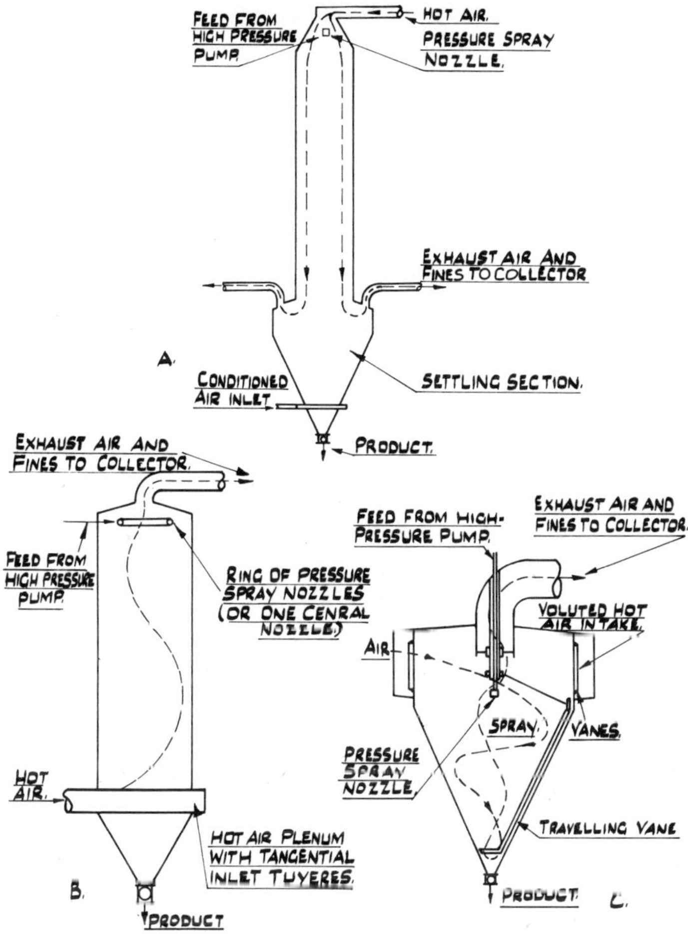

Fig. 71. Typical chamber designs for nozzle atomisation spray dryers.

Courtesy Bowen Engineering Inc.

mixed-flow spray dryer in which the hot gases enter the chamber tangentially through a series of ports, spiral downwards in the chamber and then reverse their path, leaving through a top central outlet together with product fines to the dust-recovery equipment. The main product outlet is, however, at the base of a conical bottom. Some designs are fitted with an internally driven scraper vane which keeps the internal surface of the drying chamber swept clean of material deposited on the chamber wall. One advantage of the mixed-flow design is that it can result in higher product bulk density and lower dust loading on the dust-recovery equipment. Due to the higher gas mass flow rates in contact with the chamber walls, the temperature of the latter may be higher than in other systems and as a result inlet gas temperatures may have to be limited. Furthermore, this arrangement may mean that the chamber has to be operated under a positive pressure and dust losses due to leakage to the outside atmosphere may be a drawback.

Centrifugal atomiser spray dryers generally take the form shown diagrammatically in Fig. 72 and it will be noted that they are usually of the co-current type. This arrangement is the more usual in commercial installations as it permits simpler design and more general application. Due to the almost instantaneous evaporation of moisture, the hot inlet gases are cooled in a very short distance almost to their outlet temperatures, and as a

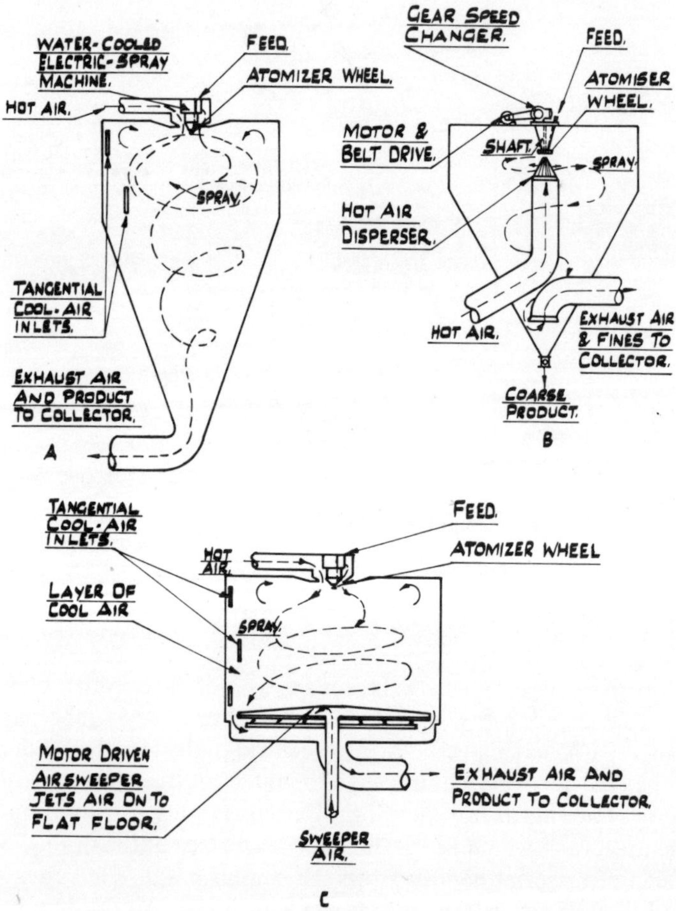

Fig. 72. Typical chamber designs for centrifugal atomisation spray dryer.

Courtesy Bowen Engineering Inc.

result the chamber wall temperature is kept to a minimum. Product temperature is invariably below the gas outlet temperature and yet very low final moisture contents of the product can be attained.

Fig. 72(*a*) shows a co-current turbulent spiral flow system, the main hot gas supply being introduced centrally around the atomising unit and a secondary tangential gas supply, either at the temperature of the main supply or at a lower level, being introduced in the upper section of the chamber, thus producing a mixed-flow pattern to the movement of gases. This tangential introduction is usually in a direction opposite to the effect of the centrifugal atomiser, thus creating turbulence and resulting in improved drying effects. Smaller chamber diameters are possible at the cost of increased height in order to accommodate a conical base to promote a cyclonic effect for dry material discharge at the bottom outlet. Fig. 72(*b*) is another co-current arrangement in which the hot gas stream is introduced below the atomiser through a disperser, the exhaust air leaving from the conical base of the dryer. In some designs the hot gas inlet disperser can be adjusted to direct the air stream at will relative to the spray movement. Generally this arrangement calls for larger chamber diameters than height, conducive to long retention times and a lesser degree of internal turbulence. In this design also the tendency is towards a spiral air flow. Fig. 72(*c*) shows a flat-bottomed chamber design provided with an auxiliary air-sweeper system, mechanically driven, which due to the effects of relatively high velocity air jets playing on the chamber flat bottom keeps the latter free from settled product which is entrained in the exit gases. In some designs a tangential supply of cooler gases is introduced along the chamber wall through ports with a view to maintaining the chamber walls at a low temperature in order to deal with low-temperature melting or softening products. The dried air-borne product particles may also be cooled by this auxiliary gas or air supply where necessary. This flat-bottomed design lends itself to internal tiling or other means of protection against corrosion without the use of expensive corrosion-resisting alloys.

In view of the considerable exposed surface areas of spray-drying chambers it is important that they should be adequately insulated on their external surfaces, the quality and thickness of insulation depending upon the anticipated surface temperature. The hot gas inlet equipment and ducting should also be insulated to restrict heat losses leading to low thermal efficiency. Outlet ducting should be insulated to the extent necessary to keep the exhaust gases above their dew-point and the same applies to dust recovery equipment such as cyclones and bag filters.

Product collection may be carried out in various ways. Where a considerable amount of product separates out within the dryer chamber in the conical base, it may be removed continuously under its own weight through a rotary valve or may be led away by means of a screw conveyor. Where most of the product remains entrained in the gas stream, and in any event, where entrainment occurs, which is usual, separation of the dry material is first carried out in high-efficiency cyclones, followed where necessary by bag filters or wet scrubbers. As cyclones of good design usually have a recovery of only 98–99%, and even a 1 or 2% loss would usually represent a considerable monetary loss, it is usually essential to follow the cyclone or cyclones with a more efficient means of recovery, even if, as with wet scrubbers, the recovered product has to be reprocessed. The use of several small cyclones or multi-clones is not unusual in order to attain the maximum product recovery in a dry state. The cyclones may, of course, be located away from the drying plant and may constitute a pneumatic conveying system to storage bins or a bagging plant. In view of the pressure losses incurred by cyclones and filter-bag units, which may amount to several inches water gauge, extraction fans are usually located after the drying chamber to avoid positive pressures therein, and most frequently after the cyclones to reduce the amount of solids handled by the fan. Wet scrubbers often employ the initial feedstock as scrubbing agent. This has the advantage where exhaust gas temperatures are relatively high in improving the thermal efficiency of the operation and in some cases preconcentrating the feedstock.

Spray dryers, often of considerable capacities, are intended to operate continuously with the least possible supervision. Automatically operating instrumentation is therefore essential, both for reasons of safety and for uniform operation. All the usual automatic flame safety devices must be included on direct and indirect fuel-fired heaters and complete interlocking of safety and operating controls for start-up and shut-down procedures, particularly in the case of the latter to guard against accidental failure of any part of the system. Operating controls are usually interlocked with the exhaust gas temperature controller which will actuate the feedstock feed-rate controller and the fuel or heat input controller.

Although a great deal has been published on academic aspects of spray drying, determination of the spray drying characteristics of a given material and the appropriate design and sizing of a commercial unit has to rely on laboratory large pilot-scale tests and on the engineering experience in this field of the dryer manufacturer. Preliminary tests may be carried out in quite small units of, say, 2–3 ft in diameter, but data reliable enough for scale-up to large industrial plants should preferably be of not less than 7 ft in diameter and should be available in different forms as regards chamber configuration, atomisation systems and air-flow patterns. The small-scale unit will establish whether the feedstock can be spray-dried and will indicate operating conditions for obtaining a suitable product including temperature, drying rate, the most suitable method of atomisation and possible difficulties likely to be met with in the atomiser. Samples of dried product can be obtained for examination both by the designer and customer. Such tests, however, may not be able to establish the conditions required for optimum particle size and bulk density and will usually set a limit on the feedstock concentration unless means for a pumped or pressure supply is incorporated. Final and reliable data from which to design the commercial unit and establish optimum operating conditions for maximum thermal efficiency and the desired product quality will be obtained on the large-scale pilot plant. Test runs of several hours would be carried out to stabilise conditions and perhaps to prepare enough product to test a market or process application. They will also establish the optimum relationship between feed rate and atomiser speeds in centrifugal atomisers, as regards product size and bulk density, or nozzle pressures using other types of atomiser. Thermal requirement evaluations can be obtained from which the economics of the commercial unit may be assessed, and a material balance can be drawn up between feed rate and product recovery from the main chamber and dust-recovery units.

Preliminary tests carried out in the small-scale pilot unit will study the effect of different feed rates against fixed inlet temperature and different inlet temperatures against fixed feed rates on product quality. Where any question of feedstock concentration arises, the effect of this will be studied to ascertain the optimum concentration commensurate with a satisfactory product. From these, the optimum combination of maximum inlet temperature, maximum feed concentration and rate to yield a satisfactory product will be arrived at, which – together with the data obtained on the most suitable method of atomisation – will set the probable conditions for large-scale pilot-plant runs. These will determine the most satisfactory combination of chamber configuration and operating data from which to design the full-scale plant.

For the design of a commercial unit the following data will have to be used:

Feedstock	Percentage solids	
	Feed rate, lb/h	
	Temperature, °F	
Product	Percentage moisture	
	Production, lb/h	
	Temperature, °F	
Evaporation rate	Moisture removed, lb/h	
Air or gases	(1) Ambient air,	dry-bulb °F
		humidity lb/lb

	(2) Entering dryer,	dry-bulb °F
		humidity lb/lb
	(3) Leaving dryer,	dry-bulb °F
		wet-bulb °F
		humidity lb/lb
Fuel	Calorific value, Btu/lb	
	Hydrogen content, % wt	
	Consumption, lb/h	

The quantity of air required to carry out the necessary evaporation is calculated, as for pneumatic dryers, by use of psychrometric and enthalpy charts, allowance being made for probable heat losses – generally known to dryer manufacturers for different sizes of dryers from experience. If combustion of fuel is used to produce the hot drying gases, the moisture arising from combustion has to be taken into consideration and this will be a function of the hydrogen content of the fuel and the temperature to which the gases are heated. Fig. 73 gives curves showing the relationship between temperature and humidity for methane, natural gas and a typical range of fuel oils. For a given hot gas inlet temperature the corresponding moisture content in lb/lb of dry gas can be determined assuming ambient air at 70°F and 0·01 lb/lb humidity. A suitable correction would have to be made for ambient air at any other condition. In the case of indirect-heated air no increase in humidity occurs. The enthalpy of this heated air will be reduced by an amount dependent on heat losses from the heated gas duct and the drying chamber and should be adjusted by an amount depending

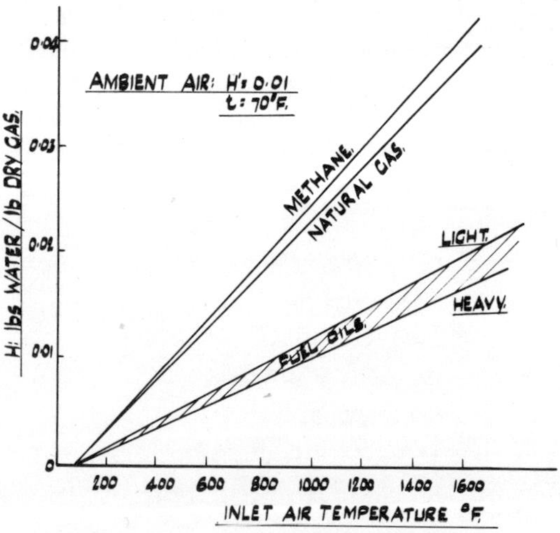

Fig. 73. Relationship between inlet air temperatures and combustion gas humidities.

on the temperature level, exposed areas and thickness and quality of insulation. This will yield the enthalpy of the exit gases at the determined exit gas temperature, whereas from a psychrometric chart the humidity and specific volume of the exit gases is obtained. The capacity of the gases for picking up moisture is obtained by difference and, the total evaporational requirements being known, the quantity of air required can be determined. The size of dryer (chamber volume) is a function of total moisture evaporation, the rate of evaporation at the stated inlet temperature and the gas feed rate. Spray-dryer manufacturers use suitable factors based on experience from which to compute the drying chamber size. In the absence of knowledge of such factors, a very approximate indication of the dryer chamber volume can be arrived at from the quantity of inlet gases, their specific volume at the outlet temperature and the assumption that the residence time for the gases within the chamber will lie between 4 and 10 seconds for chambers up to 15 ft in diameter and 10–20 seconds for larger diameters. The ratio of height to diameter will depend on several factors, feed rate, method of atomising, nature of feed and product and will be dictated by pilot-scale tests and experience.

The performance and efficiency of spray-drying plants have materially improved over recent years with the increased experience and engineering expertise possessed by dryer manufacturers. Permissible higher inlet gas temperatures have resulted in increased efficiencies and decreased costs. Fig. 74 shows a typical relationship between inlet air temperatures and heat

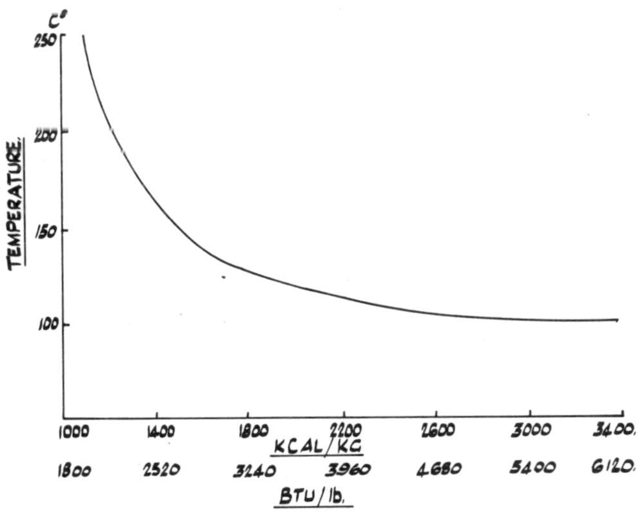

Fig. 74. Relationship between inlet air temperatures and heat consumption per kilogram of water evaporated. *Courtesy Anhydro A.S.*

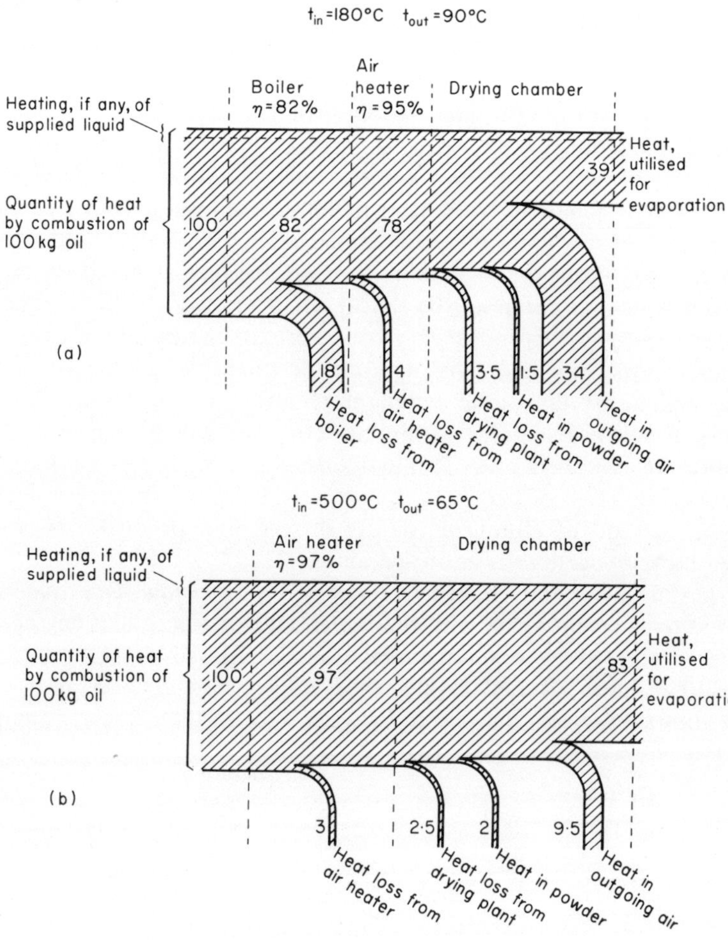

Fig. 75. Typical heat balances on indirect and direct oil fired spray dryers. *Courtesy Anhydro A.S.*

consumption met with in spray dryers whilst Fig. 75 shows two typical heat balances on a steam-heated and a direct oil-fired spray dryer. The overall improvement in thermal efficiency in the case of the direct-fired, high inlet temperature dryer over the indirect steam-heated hot air dryer is very apparent. Whereas with the steam-heated dryer the overall efficiency is 39–40%, direct fuel-fired dryers may give a heat utilisation of 80–85% for inlet temperatures of the order of 500–550°C and low outgoing temperatures of around 65–70°C. In the former, the heat losses before the drying medium reaches the drying chamber may be as high as 20–25%, whereas with direct-fired air heaters the loss is extremely low, being of the order of 3–5%. Neglecting the inefficiency of the boiler plant and steam transfer line, the steam consumption of spray dryers may be 1·2–1·8 lb of steam per pound of moisture evaporated.

In view of the considerable quantities of moisture which are removed in a spray dryer by thermal means, together with the relatively high capital investment in plant—especially where stainless steel is used for contact surfaces—an indication of the probable cost of spray drying a product is valuable. This will depend largely on the amount of moisture to be removed and the ease with which a particular product releases the moisture. It will also depend on the method of heating, the inlet temperature and the outlet temperature. The heat sensitivity of the material will also have a bearing. The ease with which the product parts with its moisture, as also its heat sensitivity, is empirically related by some dryer manufacturers to a factor known as the product number. This number increases with increased difficulty in drying or in heat sensitivity or both and is arrived at from pilot-scale test data or previous experience. Product numbers for a few typical classes of materials are given in Table 34.

Table 34. Product numbers for a few typical materials.

Material	*Product number*
1. COLOURS	
Reactive dyes	5– 6
Pigments	5–11
Dispersed dyes	16–26
2. FOODSTUFFS	
Carbohydrates	14–20
Milk	17
Proteins	16–28
3. PHARMACEUTICALS	
Blood insoluble/soluble	11–22
Hydroxide gels	6–10
Riboflavin	15
Tannin	16–20
4. RESINS	
Acrylics	10–11
Formaldehyde resin	18–28
Polystyrene	12–15
5. CERAMICS	
Alumina	11–15
Ceramic colours	10

Courtesy Bowen Engineering Inc.

A rapid estimate of overall spray-drying costs can be obtained by use of the Nomograph Chart shown in Fig. 76. The product number of the material (if not already known) is found from the intersection of the drying medium inlet and outlet temperatures as obtained on pilot-scale tests and corresponding with optimum operating conditions and product quality. Accuracy in assessing the product number is not essential as its effect on the overall cost per pound of evaporation is not great, whereas the effect of evaporational duty is considerable, the cost at 300 lb/h evaporation being five times that at 10,000 lb/h due to the greater effect of labour charges on the lower evaporation loads.

The overall cost of drying shown in Fig. 76 includes equipment costs usually encountered for a direct-fired air heater and its associated equip-

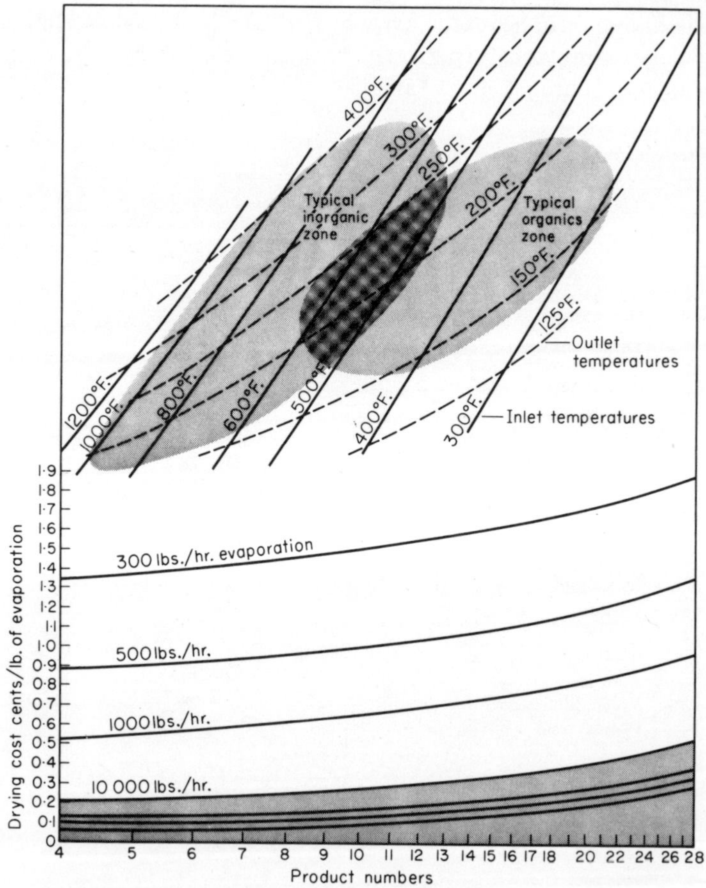

Fig. 76. Nomograph chart—showing cost of spray drying. *Courtesy Bowen Engineering Inc.*

ment, drying chamber of normal design, atomising equipment, product collector and rotary valves, main fan, interconnecting ducting, structural steel supports, mechanical drive, automatic controls and start-up charges. An average installation and erection cost is also included. Costs of special chamber design features, electric or indirect heaters and multiple product collectors are not included. C_e represents the overall cost of drying in cents per pound of evaporation. Interpretation of overall drying costs is possible by use of Fig. 76 between evaporation rates for materials of different product numbers. Thus, if tests show that with an inlet temperature of 500°F and an outlet temperature of 170°F satisfactory drying is accomplished, the product number N will be about ten. For an evaporation rate of 2,000 lb/h, the overall drying costs C_e will be 0·42 cents/lb of water evaporated.

It is more usual, however, that the cost of spray drying is required in terms of final product production, and conversion from costs per pound of evaporation to pound of product can be obtained from Fig. 77, where E/P is the ratio of evaporation rate to product rate. Thus, if in the above example the solids content of the feed were 40% and the product were dried to substantially nil residual moisture, E/P is seen to be 1·5 and the cost per pound of product will be approximately $1{\cdot}5 \times 0{\cdot}42 = 0{\cdot}63$ cents/lb. It should be noted that the above estimate of drying costs is based on labour costs at three dollars per hour, fuel costs at 50 cent/million Btu and power costs at 1·25 cents/kWh. In view of the lower labour rates and

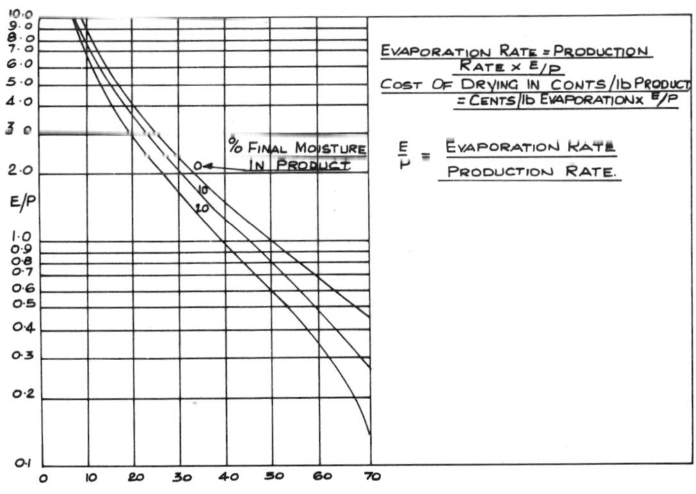

Fig. 77. Conversion of spray drying costs from evaporation to product rates.

Courtesy Bowen Engineering Inc.

high fuel costs in the United Kingdom, results estimated from the curves nevertheless give a rough approximation of the drying costs. Installation costs are estimated at 40% of the equipment cost and include no buildings; write-off is taken at 20% of equipment cost and no allowance is made for administration, sales or maintenance costs. Fig. 78 gives a curve from which the approximate oil consumption in pounds, using an efficient design of direct-fired air heater, can be calculated for different inlet air temperatures per 100 lb of water evaporated.

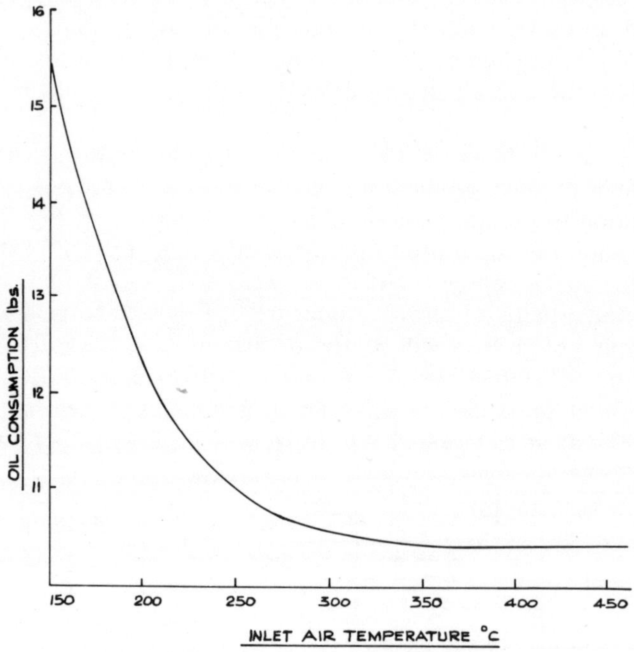

Fig. 78. Oil consumption of air heaters per 100 lb. evaporation at different air inlet temperatures.
Courtesy Anhydro A/S

As an example of sizing a spray dryer, assume that pilot-scale tests on a solution of food flavouring containing 20% of solids showed that, with a hot air inlet temperature from a steam heater of 230°F and a residence time of 6 s, a satisfactory product was obtained at an air outlet temperature of 100°F, the product containing 5% residual moisture on the wet-weight basis. Ambient air conditions were 70°F and humidity $H = 0{\cdot}008$ lb/lb dry air. Calculate the size of the spray dryer chamber required for an output of 500 lb/h of product and the capacity of the steam-heater battery.

For a production of 500 lb/h the feed would be 2,375 lb/h and the evaporation rate would be 1,875 lb/h or 31·25 lb/min. Ambient air at 70°F with $H = 0{\cdot}008$ has an enthalpy of 18·4 Btu/lb dry air and a specific volume of 13·5 from the psychrometric chart. This air, heated to 230°F with H still at 0·008 has an enthalpy of 59·5 Btu/lb and a specific volume of 17·6 ft³/lb. The exit air at 100°F will have lost some heat content due to convection and other causes; assume this amount to be 10% of the difference between the enthalpy of the inlet hot air and that of the ambient air fed to the dryer, i.e. 4·6 Btu/lb. The resultant enthalpy of the exit air will be $59{\cdot}5 - 4{\cdot}6 = 54{\cdot}9$ Btu/lb corresponding to a humidity from the psychrometric chart of 0·035 lb/lb and specific volume of 14·9 ft³/lb.

Each pound of dry air will therefore pick up $0{\cdot}035 - 0{\cdot}008 = 0{\cdot}027$ lb of moisture and the dry air requirements will therefore be

$$\frac{31{\cdot}25}{0{\cdot}027} = 1{,}154 \text{ lb/min}$$

As the pilot-plant tests showed that a residence time of 6 s was required, the volume of the spray chamber will be

$$\frac{1154 \times 14{\cdot}9 \times 6}{60} = 1685 \text{ ft}^3$$

allowing no factor of safety.

The specific volume of the ambient air being 13·5 ft³/lb dry air, the volume of air to be dealt with by the heater would be 15,580 ft³/min. The heater itself should be rated on the large side and it would be safe to allow for an outlet temperature 40° above the test-plant conditions, namely, 270°F. The enthalpy of air at $H = 0{\cdot}008$ and at 270°F is 70 Btu/lb dry air, so that the heater should be designed for a maximum heat transfer at the available steam pressure of

$$1{,}154 \times 70 \times 60 = 4{,}800{,}000 \text{ Btu/h}$$

whereas the anticipated normal operational requirement would be

$$1{,}154 \times (59{\cdot}5 - 18{\cdot}4) \times 60 = 2{,}850{,}000 \text{ Btu/h}$$

CHAPTER XI

Film Drum Dryers

Film drum dryers, sometimes referred to as roller dryers, are essentially conduction dryers, the drying effect being obtained by the transfer of heat from a heating medium within a rotating cylinder through the metal of its construction to a film of liquid containing a recoverable solid in suspension or solution formed on the external ground or polished surface of the cylinder. Convection effects only apply to the extent that ambient air convection currents occur over the cylinder surface or, in special designs, where a forced convection air current is applied to this surface. Film drum dryers may be designed with either one or two drying cylinders, mounted on a horizontal axis and mechanically rotated, usually with variable speed control. They are used for drying solutions, slurries and various solid suspensions usually in water and certain gel-like liquid products from which the solids content has to be recovered in a dry or near-dry condition. As film dryers are operated as a rule at rotational speeds varying from 5–20 rev/min, the time of contact of the material with the heated surface of the cylinder is short, say, from 3 to 12 s. As free moisture is present for a substantial portion of this period, the solids content attains the cylinder surface temperature for an exceedingly short time. Film drum dryers are therefore suitable for most materials with little or no thermal degradation in the course of drying, and where the material is unduly sensitive to temperature special precautions may be taken in the design to counter this effect.

The capacity of a film drum dryer will depend upon the effective drum surface in contact with the feedstock, the drying characteristics of the material, the temperature of the heating medium and thickness of the cylinder wall, the solids content of the feedstock and the uniformity and thickness of the film which is or can be applied to the rotating cylinder. It will also depend upon the rate of heat transfer obtainable with the heating medium employed; in practice this is usually condensing steam, although in somewhat rare instances hot liquid heating media, radiant heat elements and hot combustion gases are used. In general, it may be said that film drum dryers are employed for drying operations suitable for spray dryers but the physical nature of the product, scale of operations, site considerations or economics may have an important bearing on the selected type of dryer. A spheroidal free-flowing product, and one which retains the original flavour or aroma properties, large tonnage productions and installation in the open will usually dictate the use of a spray dryer. Medium and small productions of a variety of products, low-cost steam or a demand for low pressure exhaust process steam and low-price labour may indicate a preference for film drum dryers.

Film drum dryers are of either atmospheric or vacuum design, more usually the former. If the latter, the film dryer is housed in a vacuum chamber, a continuous flow of feedstock being introduced whilst the dried product is continuously discharged, usually to two receivers operated in parallel. In either case the feedstock is applied to the effective drying surface of the dryer by one of several methods and the dried product is removed continuously by the scraping effect of a stationary knife arrangement, which may be in the form of a single knife blade or of a number of knife blades abutting on or overlapping one another, with provision for adjusting the pressure of the knife on the cylinder surface or its degree of clearance. Drum dryers take the form of single-drum, double-drum or twin-drum designs. The single-drum dryer, as its name implies, consists of only one drying cylinder and one scraper (or "doctor") knife assembly. The twin-drum arrangement is composed of two single-drum dryers with a common drive assembly, possibly but not always two separate feed arrangements and has two scraper knife assemblies. The direction of rotation of the drums and the location of the scraper knives depends on the method of feed employed. The double-drum dryer has two drying cylinders with a common drive and common feed arrangement, the latter being into a trough formed above the "nip" between the two cylinders by the use of spring-loaded end plates or "cheeks" bearing on the machined flat ends of the cylinders and built of or faced with material suitable for contact with the liquid feedstock and having suitable heat- and wear-resistant properties. The direction of rotation of the cylinders is almost invariably downwards towards the nip between the cylinders, the clearance at which point is adjustable by adjusting the position of one cylinder relative to the other. Typical diagrams of these three types, with different feed arrangements are shown in Fig. 79.

A typical single-drum dryer is illustrated in Plate XI. As a twin-drum dryer is essentially two single-drum dryers mounted in a common frame and having a common drive, the different design features of the one are common to the other. Two major advantages of these types are that an

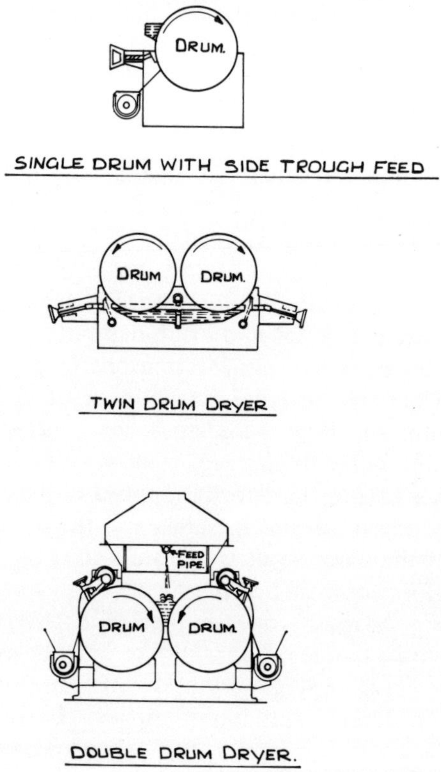

Fig. 79. Single, twin and double drum dryers.

absolute maximum of effective drying surface is possible due to the close proximity of the scraper knives to the feed arrangement and the elimination of a nip requiring adjustment between two cylinders. An effective surface amounting to 85–90% of the total cylinder circumferential area is possible as against the more usual 60–65% in a double-drum dryer. Film thickness control is not as a result of adjusting the nip between two cylinders as in the double-drum dryer where a limitation is set by the necessity for preventing fall-through of feedstock at the nip. Furthermore, there is always the possibility of damage to the cylinder surfaces of a double-drum dryer in the event of foreign bodies such as nuts, bolts or spanners being introduced. A further advantage of the single- and twin-drum dryers over the double-drum dryers is that in the former the major moisture vapour release is out of contact with the surfaces of largely dried product and the discharge of fully dried material at the knives. In the case of dusty material and material which is discharged at the knives as a fine powder or very light flakes, the dust nuisance is thereby reduced, and drying may be more rapid due to the lower humidity of the atmosphere present in the later stage of drying. The double-drum dryer may, on the other hand, sometimes prove advantageous due to the amount of pre-heating (and often even boiling and consequent pre-concentration) of the feedstock present in the trough over the nip between the cylinders.

The possible feed arrangements on single- and twin-drum dryers are many, and are illustrated diagrammatically in Fig. 80.

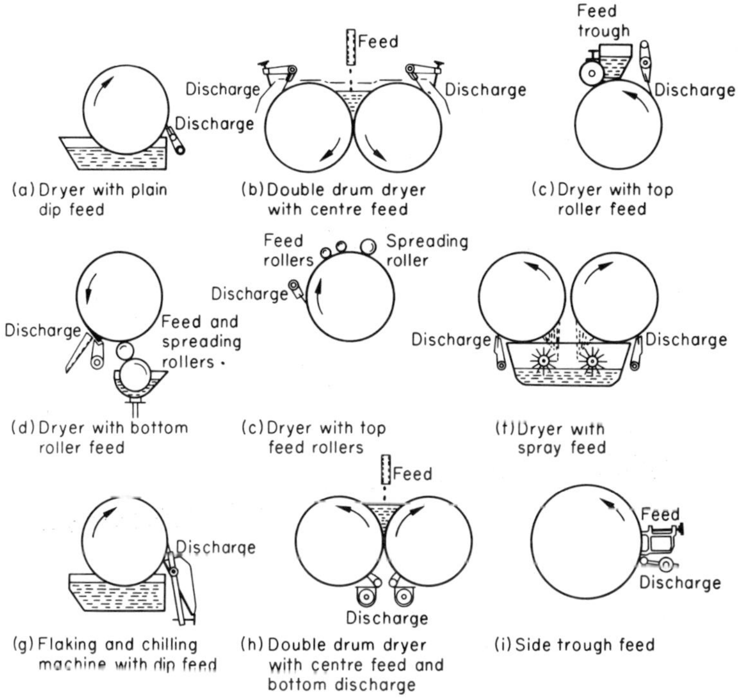

Fig. 80. Film drum dryer feed arrangements. *Courtesy A.P.V.-Mitchell (Dryers) Ltd.*

The simplest arrangement, Fig. 80(*a*), applicable equally to both single- and twin-drum dryers, is the bottom trough feed. The trough may be jacketed for either heating or cooling, is sometimes provided with an agitator in the case of suspensions or may be provided with a recirculation system to prevent settling out of material. Provision is usually made for adjusting the height of the trough, the level of its contents being maintained constant. The depth of immersion of the drum surface, together with the speed of rotation, controls the film thickness. Most feedstocks other than thick suspensions and very viscous solutions can be handled by this method. It will be noted that a very high proportion of the drum surface is effective for drying and that the whole of this area is free to release water vapour without danger of dry product dust entrainment. Fig. 80(*b*) shows the feed arrangement on a conventional double-drum dryer. Feedstock is introduced into the trough formed at the nip between the two drums by means of a perforated feed-pipe extending over the length of the drums, a trough with a serrated weir, or by a swinging pipe mounted above the drums. The liquid at the nip is frequently heated to boiling point and a degree of pre-concentration may be obtained which will generally improve the capacity of the dryer. This pre-heating, however, may not be permissible with some materials and for heat-sensitive products this system may not be satisfactory. The double-drum dryer is, however, very widely used as its manufacture is highly standardised, with consequent reasonable first cost. Where the level of liquid at the nip has to be kept extremely low because of danger of product deterioration, the effective drum surface is lower than with single- and twin-drum dryers. Fig. 80(*h*) shows a double-drum dryer arranged with the doctor knives located below the nip and with the drums rotating upwards at the nip. This arrangement is unusual and has the disadvantage that the clearance between the drums must be minimal as otherwise feedstock tends to drip on to the dry product: control of film thickness is therefore very difficult and will depend on the speed of rotation of the drums and the viscosity of the feedstock.

Figs. 80(*c*), (*d*), (*e*) and (*i*) show a variety of feeding arrangements frequently used on single-, and sometimes on twin-drum dryers. Arrangement (*c*) is sometimes used as an alternative to the bottom trough feed where pre-concentration is desirable, whereas arrangements (*d*) and (*e*) are used where this is disadvantageous. In system (*d*) feedstock is picked up from a feed trough by a roller which may be cooled and is transferred to the drying drum by means of a second feeder-roller whose speed of rotation can be controlled to determine the film thickness on the drying drum. This arrangement is particularly suited to drying heat-sensitive viscous liquids. Fig. 80(*e*) shows a series of feeder and spreader rolls, some or all of which may be either heated or cooled and between the first two of which the feedstock is introduced. Such a dryer is also illustrated in Plate XXXI. Relatively close control is possible of the degree of pre-heating of feedstock and the method can therefore be used for pre-cooking or pre-digestion of the feedstock as for infant cereal foods or the pre-gelatinising of starch solutions. Sometimes additional feeder and spreader rolls are used to build up a substantial film thickness. This system can be equally applied to twin-drum units and has been used extensively for potato- and maize-flour production. The feed and spreader rolls are usually independently driven at controllable speeds or may be gear-driven from the main-drum drive with interchangeable gear wheels for different speeds. Fig. 80(*f*) shows a spray or splash feed applied to a twin-drum dryer in which the feedstock is splashed against the drum

Plate XXXI. Top-roller feed film dryer. *Courtesy A.P.V.-Mitchell (Dryers) Ltd.*

surface by means of rotating paddles or brushes. Such liquid as does not adhere to the drying drum falls back into the feed trough resulting in a degree of pre-concentration. Film thickness is not amenable to close control and the system is quite unsuitable for liquids which suffer frothing on agitation or aeration. With all types of feeding arrangements the objective is to produce as even a distribution of liquid as possible over the full length of the drying drum, and for this reason precautions must be taken to ensure a truly horizontal level to the liquid surface in dip troughs or correct aligning of applicator rolls. Multiple liquid inlets and offtakes in dip troughs and adjustable serrated weirs in side troughs are an advantage. Because of the possibility of over pre-concentration recirculation on feed troughs is also resorted to. It is always advisable to arrange that feed troughs can be completely drained at the end of a run and that applicator rolls can be washed down.

Doctor knives are generally one of three types: stationary single-bladed; oscillating single-bladed; and multiple adjustable abutting or overlapping blades. For materials which form a film of dried material which is easily removed by the knife, a single knife extending the whole length of the drying drum, provided with adjusting screws at frequent intervals and the whole knife assembly or beam adjustable for clearance or bearing pressure, is satisfactory and inexpensive. The knife will usually be of spring steel of fairly small thickness set at an acute angle to the tangent at the point of contact. Sectional knives will usually be of more robust thickness and set at an angle most suited to the material film. Each section is adjustable by means of double hand wheels, one of which is capable of very fine adjustment to knife clearance. For materials which are extremely abrasive the knives are often tipped with abrasion-resistant metal; where corrosion is likely to occur or where metal contamination has to be completely eliminated the knives may be of a suitable plastic material or even toughened glass. A sheet metal or stainless steel chute plate is usually fitted to deflect the dried product away from knife adjustment gear where this is located below the knife assembly.

It is usual to provide drum dryers with a means of variable speed control over the rotational speed range of the drying drum likely to be experienced. Final drives may be chain, spur-gear wheel and pinion or a worm-wheel drive, usually manufactured in phosphor bronze. All gear wheels should be preferably machine cut. Drying drum speeds may vary over the range of 3–20 rev/min depending on the design and the nature and drying characteristics of the feedstock, usually ascertained during pilot-scale tests.

Drum dryers must be the subject of exhaustive pilot-scale testing before a choice is made of this design of dryer, its type, size and method of feed. In the first instance it must be established whether the feedstock will in fact form an adequate and uniform film on the drying drum by any method of feed. Secondly, such a film may be produceable on one material of drum construction but not on another. Whereas for normal purposes a close-grained cast iron, possibly with an appreciable nickel content, is usual, possibility of corrosion or contamination may dictate the use of non-ferrous or special alloy or plated drums, and it is essential to establish that the feedstock will form a suitable film on such surfaces. A method of feed must be chosen that will give a good and even film distribution without deterioration of the feedstock due to temperature rise or settlement of material, the desideratum being the production of a product of composition identical to that of the solids content of the feedstock. Steam pressure within the drum may be critical, as excessive steam temperature may cause loss of film-covering capacity due to blistering of the feed on the drum. The peripheral speed of the drum may have an effect on film pick-up and thickness as important as adjustment to the feeding arrangements. Excessive abrasion or metal pick-up may mean discarding this type of dryer whilst undue dust formation or a product of unsuitable physical characteristics may have the same effect. Once it has been established that a film drum dryer is suitable and the most satisfactory method of feed has been ascertained, the production rate per square foot of effective drying surface can be determined. Regular and uniform operation of the full-scale dryer will usually give production rates at least equal to that obtained under optimum small-scale tests so that the sizing of the dryer is straightforward, a factor of safety being allowed to give adequate flexibility for shut-down periods or market demands.

Drying of liquids on film-drum dryers involves (in the case of steam-heated dryers) the transfer of heat from condensing steam within the drum through the metal thickness of the drum itself to the thin layer of feedstock on the outer surface of the drum, the composition (moisture content) and temperature of which is progressively changing from feed point to discharge point. The overall heat transfer rate as drying proceeds will therefore vary considerably and will depend upon:

1. Solids content and physical properties of the solids.
2. Steam temperature.
3. Film thickness.
4. Metal shell thickness and its thermal conductivity.

Neither the solids content nor the film temperature remains constant during the drying operation so that determination of overall heat transfer rates can only be made on a pilot-scale unit. As drying proceeds the conductivity of the product becomes important; whereas the temperature difference between the feedstock and the steam-heating medium will be a maximum at the feed point this may fall to a very few degrees at the discharge point. The heat-transfer coefficient on the steam side is not predominant except in cases of extremely high evaporational rates, although it is most important to provide adequately for bleeding off incondensable gases and condensate. Metal thickness, however, assumes considerable importance and should be maintained at a minimum commensurate with satisfying sound design and insurance requirements. In the case of very concentrated solutions or suspensions, doubling the drum wall thickness may result in almost halving the dry product production rate. Work carried out by Van Marl[(14)] indicated that on a double-drum dryer the overall heat-transfer coefficient under optimum conditions varied from 360 Btu/h/ft^2/deg F at the nip to about 220 Btu/h/ft^2/deg F in the area adjacent to the knives, the coefficients on the steam side and liquid side ranging from 2,650–450 Btu/h/ft^2/deg F and 1,500–1,800 Btu/h/ft^2/deg F, respectively. The temperature drop through the metal drum ranged from 65–35°F. Under favourable commercial conditions the maximum evaporation rates normally encountered are of the order of 18·5 lb/h/ft^2 drum surface, but evaporation rates much lower than this are not unusual. This unknown factor of heat-transfer rates and the effect of drum speed and film thickness makes pilot-scale tests imperative in sizing a film dryer.

Thermal efficiences of film drum dryers, assuming adequate insulation of the drum ends, are high, being of the order of 80–90% corresponding to a specific steam ratio of from 1·2–1·5 lb of steam per pound of moisture evaporated. Introduction of air heated by the condensate over the film surface by the use of a close-fitting cowl surrounding the drum-drying surface may in some cases give even higher efficiencies. The thermal efficiency of a drum dryer installation will naturally depend upon the design type, as the greater the ineffective drum area the greater the heat loss. Single- and twin-drum dryers consequently operate as a rule at higher efficiences than double-drum dryers. Very low final moisture contents of product may also reduce efficiencies.

Optimum capacities of film drum dryers generally result from the most favourable combination of solids concentration, drum speeds and heating medium temperatures to suit a particular feedstock. Capacities increase with increased solids content; usually increase with steam temperatures; and increase with drum speeds over the lower ranges. Above a certain speed of rotation film thickness usually decreases, leading to loss of capacity unless compensated for by increase in steam temperature. Beyond certain steam temperatures, on the other hand, blistering effects with some feedstocks will result in loss of production.

Vacuum film drum dryers, which are designed generally on the same lines as atmospheric drum dryers but are housed in vacuum-tight chambers, are used for materials which are particularly heat-sensitive, are subject to oxidation or where recoverable solvents are the component which has to be evaporated. Such a dryer arranged for bottom trough feed is shown in Fig. 81, the dryer itself being of the single-drum type.

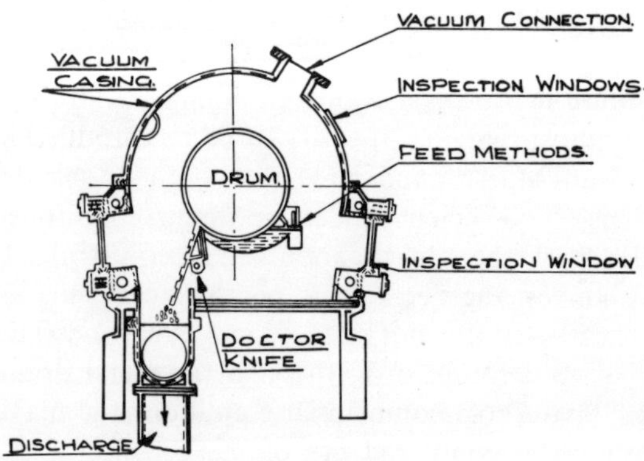

Fig. 81. Single drum vacuum film drum dryer.

Double-drum dryers are also used under vacuum but suffer from design complications as compared with single-drum units in view of the special requirements for drum to drum adjustment at the nip. Feed arrangements can be generally similar to those employed on atmospheric units, but the dry material discharge is usually through a screw conveyor leading to a common discharge pipe to twin receptacles maintained under vacuum and operated intermittently in parallel. Feed control, knife adjustment, spreader rolls and, in double-drum dryers, the clearance at the nip have to be controlled from without the vacuum casing using vacuum-tight stuffing boxes and glands. It is usual to provide observation windows so that visual control of adjustments can be made, and the vacuum casing should be strip- or tracer-heated to reduce recondensation on the inner surfaces of the casing to a minimum. For materials which are particularly heat-sensitive it may be necessary not only to have a reduced pressure atmosphere surrounding the drying drum, but also to supply steam to the drum at a sub-normal pressure. Such an arrangement, compared with supplying a liquid heat transfer medium, has the advantage of still retaining the relatively high heat-transfer coefficient inherent in the use of condensing steam. Table 35 gives the temperature of steam at specified pressures below atmospheric together with the heat contents of water and steam at these pressures and temperatures.

Table 35. Properties of dry saturated steam at depressed pressures

Pressure (in) vacuum	*Temperature of boiling* °*F*	*Heat in water Btu/lb*	*Latent heat of steam Btu/lb*	*Total heat Btu/lb*
29·0	77	45	1,050	1,095
27·0	114	82	1,029	1,111
25·0	133	101	1,018	1,119
20·0	161	129	1,002	1,131
15·0	179	147	991	1,138
10·0	192	160	983	1,143
5·0	203	171	976	1,147
1 atm	212	180	970	1,150

Thus, if steam is supplied to the drying drum and maintained at a pressure of 25 in vacuum within the drum, its temperature will be 133°F, below which it will condense, releasing 1,018 Btu/lb, whereas water supplied at that temperature would only release 1 Btu/lb for each degree through which it is cooled. As the water temperature drop can only be a few degrees in order to maintain an adequate drying temperature, a very considerable quantity of hot water will have to be circulated to release sufficient heat to produce evaporation of moisture. Furthermore, in order to secure a reasonably high heat-transfer rate from the water to the drum shell, the internal design of the latter will need to be such as to give high water velocities in contact with the shell. Alternatively, internal high-velocity water jets may be used impinging on to the internal drum surfaces to break down the water film.

The degree of vacuum which should be maintained within the casing surrounding the drum dryer will depend largely upon the material being dried but will usually be maintained as high as conveniently possible. Fig. 82 gives a curve showing the vapour pressure of water plotted against temperature. Vapour pressures of solutions and fine suspensions will not conform strictly to figures derived from this curve, but it is a useful indication of the effect of evaporating water from a film drum dryer under vacuum. Thus, at 133°F, pure water exerts a vapour pressure of about 125 mm of mercury (5·91 in) so that it would boil at a vacuum of 29·92 − 5·91 or 24·01 in of mercury.

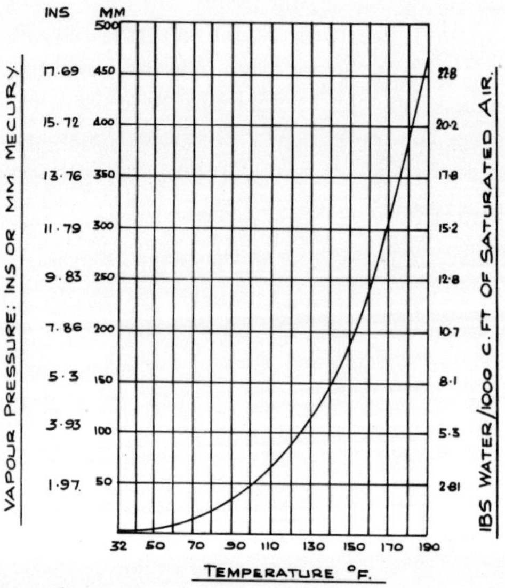

Fig. 82. Vapour pressure of water.

With a combination of 24 in vacuum steam within the drum and the same or higher degree of vacuum within the casing, conditions would be such that the aqueous feedstock could probably be dried without raising the temperature above 133°F. Such a combination of vacuum within both the dryer drum and the surrounding casing would usually be used in cases of extreme heat-sensitivity of the product or when the original aroma or flavour of the feedstock must be retained. With all film drum dryers operated under vacuum, it is of course essential to maintain the feedstock well below its boiling point up to the stage where the film on the drum is actually formed, or in double-drum dryers to maintain the liquid in the trough formed at the nip between the drums at a low level.

Strict attention should be paid to mechanical maintenance of film drum dryers and to the correct setting of the cutting-off or doctor knives. Replacement costs of doctor knives may be high depending on the abrasiveness of the product and it is essential to have a ready supply of spare knives and, if possible, means of regrinding. Drum surfaces should be maintained in a high state of cleanliness and finish, as a certain degree of wear inevitably occurs. For this reason, single- and twin-drum units, especially if provided with multiple knife sections, have the advantage over double-drum units, as control of film thickness and prevention of dripping at the nip becomes more difficult with the latter as drum surface wear proceeds. Regrinding of drums sooner or later has to be resorted to and, where more than one or two dryers are operated, it is an advantage to standardise on size and to carry one or more spare drums as replacements. Considerable loads are thrown on the driving gear and particular attention should be paid to adequate and correct lubrication. Overall maintenance costs on film dryers can be high and will generally be higher than on spray dryers.

Film drum dryers are used very extensively in industry and for a wide variety of products, a list of some of which is given in Table 36. Typical performance data on a few products are given in Table 37.

Table 36. Some materials dried on film drum dryers

Food industry	*Chemical industry*
Potato flakes and powder	Insecticides
Infant foods	Detergents
Milk and whey	Soap
Fruit purees	Organic sulphonates
Meat extracts	Inorganic salts
Yeast	Dyestuffs and pigments
Distillery by products	Tanning extracts
Soap mixes	Benzoates
Sago flour	
Ceramic industry	*Fine chemical and pharmaceutical industry*
Prepared slip for tiles	Calcium and barium carbonates
Stone and flint slops	Antibiotics
Ceramic colours	Glandular extracts
Glazes	Inorganic and organic salts
Insulator body	Insecticides
Zirconium silicate	D.D.T.
	Photographic chemicals

Cylinder Dryers

Whereas film drum dryers are applied to the drying of solutions, suspensions and pastes, cylinder dryers, which operate on very much the same principle, are applied to the continuous drying of a solid feedstock such as paper,

Table 37. Film drum dryer performance data

	Yeast cream	*Stone slop*	*Starch solutions*	*Glaze*	*Zirconium silicate*	*Brewers yeast*	*Clay slip*
Feed solids, % by weight	16	40	36	64	70	25	75
Product moisture, % w/w basis	5·7	0·2	5	0·2	0·2	5	9
Capacity, lb prod./h	168	420	300–400	225	1,120	146	4,000
Dryer type (a) single, (b) twin, (c) double	(a)	(a)	(a)	(a)	(a)	(a)	(a)
Drum: diameter	4 ft 0 in	2 ft 6 in	48 in	18 in	36 in	28 in	48 in
length	10 ft 0 in	5 ft 0 in	120 in	36 in	72 in	60 in	120 in
Type of feed method	Top roller	Dip	Top roller	Side	Dip	Centre nip	Side
Steam pressure, lb/in^2 gauge	80	60	80	—	80	40	40
Atmospheric or vacuum	atmos.	atmos.	atmos.	atmos.	atmos.	atmos.	atmos.
Steam consumption, lb/lb evaporated	—	—	1·3	1·3	—	—	1·35
Average effective area, %	—	—	86	—	—	—	65
Evaporation/ft^2/h	6·5	4	5	9	8·4	6	8·4

Courtesy A.P.V.-Mitchell (Dryers) Ltd.

fibre or textile piece goods and cellophane film. Although frequently of much larger dimensions, the cylinders or drums are of similar design to film drum dryer cylinders. There will usually be a large number of cylinders in a single installation, generally staggered in two horizontal rows as shown in the typical diagram in Fig. 83. The wet feedstock passes alternately over the surfaces of the upper row of cylinders and below those of the lower row, the cylinders being arranged to give a maximum cylinder area of contact with the feedstock. In special cases the cylinders may be installed in vertical rows to conserve factory floor space.

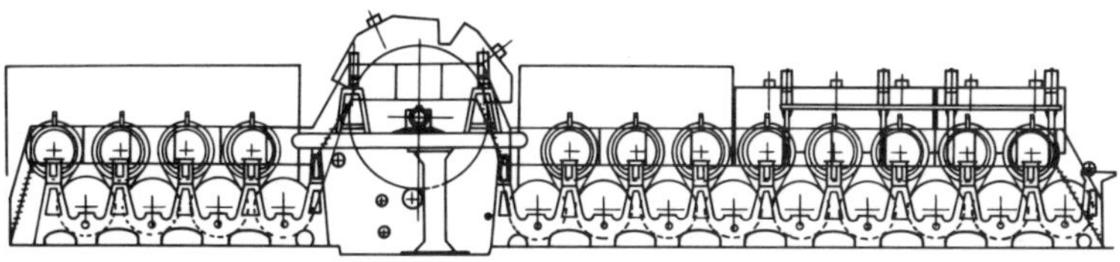

Fig. 83. Typical continuous cylinder paper dryer. *Courtesy Spooner Dryer & Engineering Co. Ltd.*

The arrangement shown in Fig. 83 brings both sides of the feedstock into contact with the heated cylinders; where only one side requires to be heated the lower row of cylinders would be replaced by unheated rolls of small diameter. As shrinkage of the feedstock usually occurs the speed of rotation of succeeding cylinders is progressively reduced. Drying is usually carried out under atmospheric conditions but in recent years forced draught enclosures over the line of cylinders have been introduced to reduce the relative humidity of the surrounding air and to promote accelerated evaporation of moisture. Still more recently, high-velocity heated air jets have been applied to the exposed paper surfaces for the same purpose.

The factors controlling the performance of cylinder dryers – ignoring the effect of external hot air jets – are similar to those for film drum dryers. Heat transfer from steam within the cylinder to the paper in contact with its external surface will be affected by adequate removal of condensate and uncondensed gases from within; cylinder wall thickness, finish and cleanliness; close contact between the feedstock and the cylinder external surface; and the temperature difference between steam and feedstock. Uniform thickness and finish, both internally and externally, of the cylinder and uniform distribution of the steam supply over the internal surface – especially with large cylinders – are essential to uniform drying of the paper sheet. Furthermore, the original feedstock requires to have uniform moisture distribution in order to ensure uniformly dried product. Heat-transfer rates of from 30–50 Btu/h/ft^2/deg F from steam to paper are normal with 70–100 lb/in^2 steam pressures and the evaporation rates run at up to 7–8 lb water/ft^2 of total cylinder surface area on newsprint machines and on other machines drying porous webs, except where high-velocity air jets are employed when they will be higher. Steam consumption is of the same order of magnitude as with film dryers, ranging usually from 1·3–2 lb steam/lb water evaporated, although it will be higher during winter time if the drying room is not air-conditioned. Output capacities of cylinder dryers can be increased by increasing the steam temperature.

In the drying of paper and similar materials on cylinder dryers, the mechanism of drying with cylinder wall temperatures up to 212°F can be explained by accepted theories of conduction and convection. Above surface temperatures of 212°F Nissen and Hansen[15] and Nissen, George and Hansen[16] have shown that the process can be explained on the basis of a series of evaporation-condensation cycles with the wet front gradually receding from the heated surface. Where felts are used, some moisture is transferred from the wet feedstock to the felt at the wet end, whilst at the same time, dependent on the degree of tension applied to the felt, the heat-transfer rate from cylinder to stock is improved with increased tension. With highly absorbent felts, the moisture transfer at the wet end of the dryer can assume important dimensions, decreasing progressively towards the dry end. Thus, on a 36-cylinder dryer processing strawpaper, Jepson[17] quotes an example of 15% loss in output of the machine by the removal of the felt off the first top six cylinders, whilst on a 23-cylinder machine processing paper for parchmentising he gives the following results illustrating the ratio of drying rate as between felted and unfelted cylinders at different positions along the length of the dryer:

	Cylinder No.					
	2 and 4	6	8	10	12	14
Ratio of drying – felted and unfelted cylinder	Felt not fitted	$3\frac{1}{2}$	3	$2\frac{1}{2}$	$1\frac{1}{2}$	1

However, it should be noted that this improvement may be as much due to the improved contact between paper and cylinder due to the tension of the felts as to the adsorptivity of the felts themselves, and recent developments have shown that the use of open-weave fabrics in place of felts gives excellent results, as they permit free flow of water vapour from the paper sheet surface in contact with the cylinder surface. Such fabrics are, of course, non-absorbent.

Forced convection drying as applied to cylinder dryers is obtained by blowing heated air at high velocity through a pattern of nozzles on to the paper or stock surface as it passes over one or more cylinders, the nozzles, heaters and fans being housed in suitably designed hoods. The result is greatly increased evaporation rates but, as such dryer accelerator hoods can only be applied in the absence of felts, the loss of drying effect of the felts themselves has to be taken into account. The net effect of the application of accelerator hoods in certain positions may not justify the capital expenditure on equipment and the power costs of operating the hoods, unless open-weave fabrics are substituted for the conventional felts, when the advantage of high air velocities is very pronounced. The whole drying mechanism down the length of the machine has to be critically examined under test conditions in order to define the variations of moisture content of the stock, the contribution of the various groups of cylinders and the effect of the felts. The result will lead to a correct decision as to how many hoods should be installed and in what positions, together with appropriate fan horsepowers and air temperatures. Finally, such increase in capacity as is secured by virtue of increased evaporation rates must be compatible with the capabilities of the installation as a whole in regard to stock preparation and treatment ahead of the drying machine.

Work carried out on an experimental paper-making machine fitted with hoods on one upper and one lower cylinder have been reported by Brauns and Larsson.[18] Sheet dryness was regulated before the hooded cylinders to simulate the effect of placing the hoods at different positions of the section of a machine and the results indicated the drying power of the hooded cylinder compared with that of a felted cylinder operating on the sheet at the same moisture content. For Kraft and newsprint, for example, it was found that the evaporation on the hooded cylinder was equal to between three and four felted cylinders at a sheet dry-solids content of 35–40%, rising to eight to ten cylinders at a dry solids content of 70–90%. Appreciably lower improved ratios would, however, be expected on production machines although the trend would be generally maintained. Jepson[17] reports that these findings agreed in principle with actual experience on large production machines. Ten accelerator hoods fitted to the top cylinders of a 21 cylinder dryer section were divided into three separate and independently controllable banks of 4, 3 and 3 hoods and the speed and drying effect of the machine running either without any hoods or in conjunction with any combination of three banks of hoods was determined. These results are reproduced in Table 38.

The effect of increased rate of evaporation due to the action of the hoods is to reduce the moisture content of the stock passing to subsequent cylinders and results in a reduction in the drying contribution of these cylinders. The contribution of the hoods in terms of extra water evaporation, as is to be expected, decreases as their application moves towards the dry end, whilst the drying contribution of the cylinders themselves ahead of the hoods improves with increased moisture content stock in contact with them. Thus, with a second bank of hoods operating, the increase in evaporation due to the hoods is 24% and that due to the cylinders is 4%, giving a net

Table 38. Effect of accelerator hoods on cylinder machines

	Hoods in operation			
	First bank of 4 hoods	*Second bank of 3 hoods*	*Third bank of 3 hoods*	*All banks of hoods*
Machine speeds No hoods = 100	115	124	134	155
Total water evaporated No hoods = 100	120	128	141	181
Extra water evaporated by hoods only: Percentage of total with no hoods	36	24	21	81
Change in water evaporated by cylinders only (loss or gain): Percentage of total with no hoods	−16 (loss)	4 (gain)	20 (gain)	0

Courtesy Spooner Dryer & Engineering Co. Ltd.

gain of 28, whereas with the third bank operating all unhooded cylinders are doing more drying, resulting in a total increase in evaporation of 41%, made up by 21% due to the hoods and 20% due to the cylinders.

Similar enhanced evaporation rates are secured from the use of accelerator hoods on machines making glazed paper, the hoods being applied to both the pre-drying cylinders and the M.G. cylinder. Jepson[(19)] quotes an example of the evaporation rate on the M.G. cylinder increasing from 7·1 lb/ft^2/h to 10·9 lb/ft^2/h (or 53%) by the use of high-velocity air jet accelerator hood with cylinder and hood-heater steam pressure of 31 lb/in^2 gauge. Steam consumption was decreased by 13·5%. On a 15 ft diameter M.G. cylinder operating with 35 lb/in^2 gauge steam and the hood heater supplied with steam at 145 lb/in^2 gauge the evaporation rate was increased to 12 lb/ft^2/h. The fan horsepower was 42 and 49, respectively. Quite considerably higher evaporation rates are possible with a combination of high steam pressure in both cylinder and heaters and high pressure two-stage axial flow fans in the hoods. Such high evaporation rates permit very high production rates at the expense of higher power consumption by the hood fans.

A combination of hooded pre-dryer cylinders and M.G. cylinder is shown in Fig. 84, accelerator hoods being fitted to both upper and lower pre-dryer cylinders. The M.G. accelerator hood is shown in more detail in Fig. 85, which shows a three-section Spooner hood, each section being self-contained with its axial flow fan, steam heater, pressure chamber and nozzles. In addition, a fresh air fan, with heater and ducting, is provided which blows the evaporational requirement of heated fresh air on to the paper immediately on its admission to the hood and contact with the cylinder. This air then passes to the suction part of the first accelerator section where, together with the recirculated air, it passes through a heater

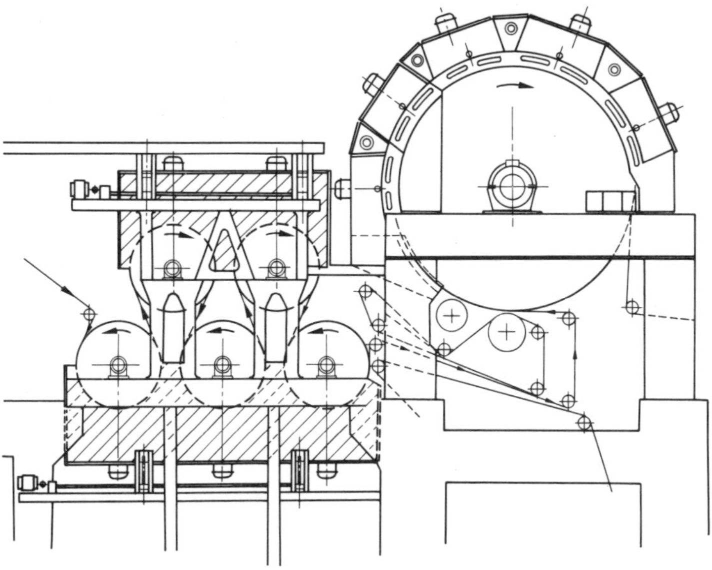

Fig. 84. Accelerator hoods to both pre-dryer and M.G. cylinders.
Courtesy Spooner Dryer & Engineering Co. Ltd.

to the section fan, which forces the air through the pressure chamber and nozzles on to the sheet. Part of the air, after being blown on the sheet, passes to the suction chamber of the next chamber along the surface of the sheet and cylinder, mixing with the circulated air in the section and being re-heated by the heater in the suction chamber of this section. This process is repeated in subsequent sections, the number of which will depend upon the cylinder size. The quantity of fresh air originally introduced finally leaves the last section together with the moisture vapour released during drying and is exhausted to atmosphere or a heat-recovery unit by an exhaust fan.

The inclusion of heaters and fans as integral components within the insulated enclosure of the accelerator hood yields a compact design conducive to minimal heat losses and absence of external trunking other than fresh air inlet and humid exhaust air ducts.

The high evaporation rate obtainable through the use of accelerator hoods depends upon air velocities leaving the nozzles, the pattern of the latter and their distance from the stock being processed and the temperature and humidity of the drying medium. Increased air velocities call for increased fan horsepowers, and increased air temperatures in steam heating require increased steam pressures. Both these factors affect operating costs and it is important to arrive at the most economical combination of the two. Fig. 86 shows the relationship between evaporation rate and electrical

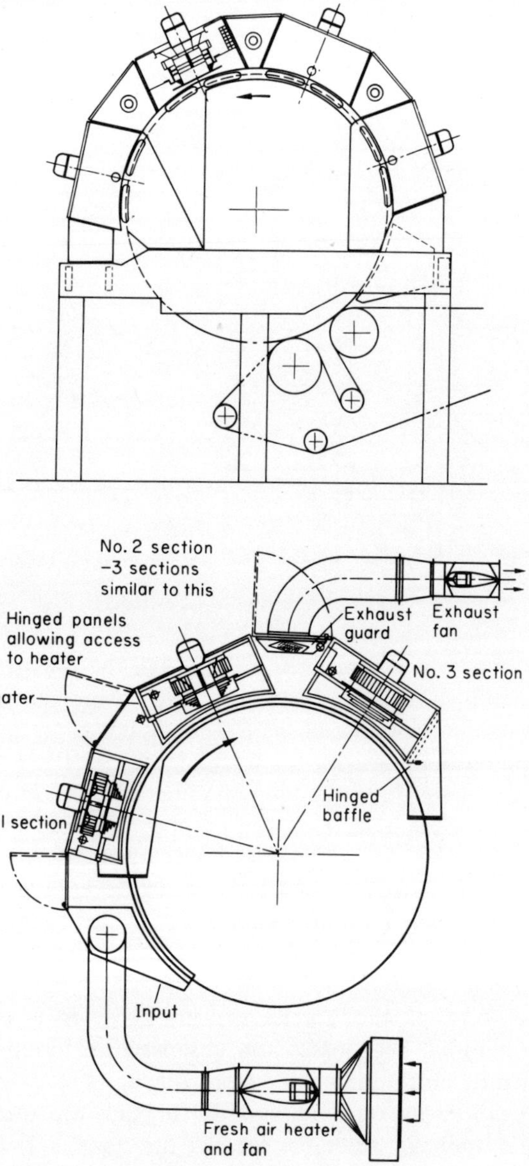

Fig. 85. Spooner "Axiflo" M.G. accelerator hood. *Courtesy Spooner Dryer & Engineering Co. Ltd.*

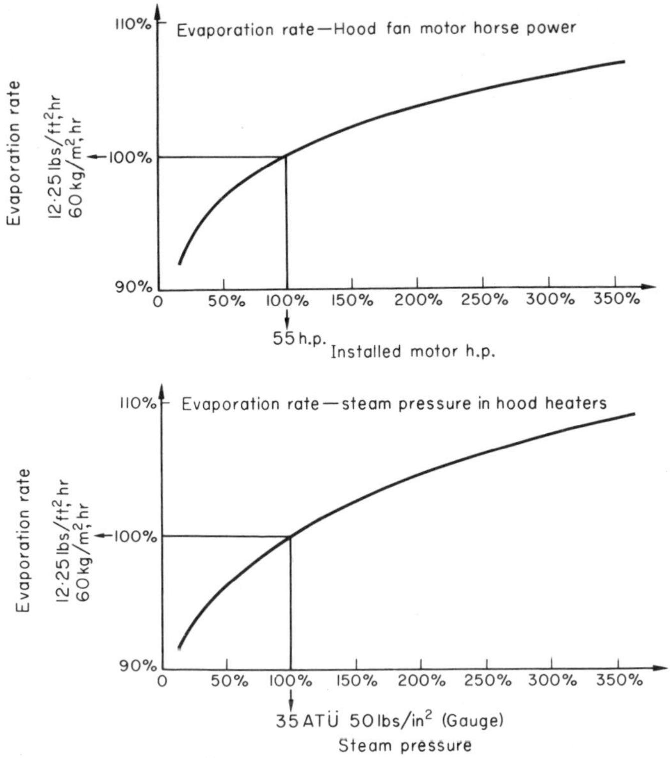

Fig. 86 Relationship between evaporation rate, fan horse power, steam pressure in accelerator hoods.
Courtesy Spooner Dryer & Engineering Co. Ltd.

power required by the fans as also the relationship between evaporation rate and steam pressure.[19]

It will be seen that in both, large increases, beyond a certain point, in either fan horsepower or steam pressures give relatively small increases in evaporation rates. The disadvantage of using high-pressure steam to secure high air temperatures may be overcome through the use of direct or indirect gas or liquid fuel heating, the high temperature being brought into contact with the highest moisture content paper entering the hood. Maximum moisture evaporation rates occur at this point with consequent rapid lowering of air temperature and minimal danger of damaging the paper. Temperature levels within the hood's air stream are such that re-heating at individual sections is unnecessary. Exhaust air is maintained at satisfactory levels of temperature and dew-point by the use of a heat-recovery system which serves heat for pre-heating fresh air introduced to the system.

Steam consumption is significantly affected by the quantity of fresh air introduced into the system and the temperature to which it is pre-heated by a heat-recovery system. Theoretically it would be possible to operate with the introduction of little or no fresh air, relying only on the drying effect of high-velocity superheated humid air. In practice, of course, it would be impossible to exclude the entry of fresh air as it would be impracticable to seal the hoods completely against the suction effect of the circulating fans. Jepson[19] gives curves illustrating the relationship between the dew-point of air exhausted from an accelerator hood, evaporation rate and steam consumption which show that the lower the dew-point, caused by increasing the quantity of fresh air introduced, the lower the steam consumption.

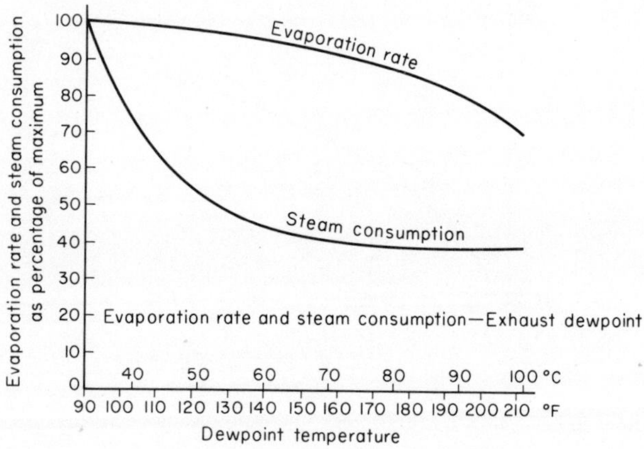

Fig. 87. Effect of exhaust air dew point on evaporation rate and steam consumption using accelerator hoods.

Courtesy Spooner Dryer & Engineering Co. Ltd.

It will be seen that, with increase in dew-point, steam consumption decreases rapidly to about 140°F but thereafter is only slightly affected, whereas evaporation rates commence to fall off materially at about the same value of dew-point. The operating point for maximum economy will depend on local steam costs and required production rates.

From the foregoing it is evident that considerable increases in the production capacities of M.G. machines and multi-cylinder drying machines are possible through the use of high-velocity accelerator hoods of good design and, in the case of multi-cylinder machines, their proper location. The replacement of felts on the latter must be located with care, not only from the standpoint of the positions at which the felts themselves assist to the greatest extent in the removal of moisture, but also having regard to other beneficial effects from the use of felts such as the maintenance of good contact between stock and cylinders, prevention of cockling, restriction of shrinkage and consequent enhancement of paper strength. In general, accelerator hoods are best applied at the earlier stages of the drying machine where maximum evaporation rates can be attained on areas of relatively high moisture content with the least danger of damaging the stock, whilst the last section cylinders would remain felted to prevent cockling. Finally, it should be noted that, as temperature has an important bearing on machine capacity, the use of direct-fired hot air supply using either gas or oil makes possible temperatures of the order of 750°F which may yield evaporation rates of 25–30 $lb/ft^2/h$.

CHAPTER XII

Indirect Agitated Dryers

The performance of indirect dryers generally, in view of the importance of good and uniform contact between the feedstock and the heated surface in order to secure optimum heat-transfer rates and temperature distribution, is materially affected by the degree of agitation of the material during the drying process. This, as has been seen, is achieved in indirect continuous rotary dryers by the repeated lifting and dropping effect of internal flights or the rolling kiln action of the rotating shell where flights are not used. Similarly in steam-tube rotary dryers this effect is obtained by the relative movement of feedstock and heated tubes. There are, in addition, a range of dryer designs in which this relative movement and change in surface contact is achieved through the use of mechanically driven agitation equipment both on a batch and on a continuous principle. Those designed to operate on the batch principle have the added advantage that where necessary they can be operated under reduced pressure for the removal and recovery of solvents, or at relatively high vacuum when heat-sensitivity of the material demands the use of low temperatures. Where the feedstock is produced by previous processes in batches it is usual to employ batch or intermittent dryers, but for large tonnage productions on a continuous basis designs of continuous agitated dryers are available which can be used if other types of dryer are not suitable.

Vacuum batch dryers, whether employing mechanical agitation or the tumbling of the charge, will usually be selected when some or all of the following requirements have to be satisfied:

1. Limiting maximum product temperature.
2. Low final moisture content at low drying temperatures.
3. Recovery of solvent.
4. Minimum loss of dust.
5. Elimination of contamination.

6. Elimination of surface "case-hardening", experienced in vacuum tray dryers.
7. Drying requires to be effected in the absence of air.

When operated under vacuum the most effective method to speed up the drying rate, which is dependent on the heat-transfer rate from the dryer heated surfaces, is to derive as much benefit as possible from heat transfer from hot vapour within the dryer in the earlier stages. Heat transfer both from this and heated surfaces is greater if the solids surfaces are still wet than when they have produced a dry skin. Hence it is an advantage to develop the vacuum within the dryer slowly on start-up, gradually increasing the vacuum pump or ejector effect as drying proceeds. In this way the solid particle surfaces remain wet with moisture migrating from the particle centres, hot vapours are present to enhance the heat-transfer effect of heated surfaces and the wet particle surfaces retain a relatively high heat-transfer coefficient. The importance of this procedure will be appreciated when one compares an approximate heat-transfer rate of from 1–2 Btu/ft^2/h/deg F between a nearly dry particle surface and the heated surface with a figure of approximately 15–20 Btu/ft^2/h/deg F between hot vapour and a moist solid surface.

Batch Dryers

Agitated Pan Dryers

When comparatively small batches of material are produced which do not lend themselves to batch fluidised bed drying and particularly if they tend to suffer from surface "case-hardening" when dried on trays, drying can be carried out with a minimum of labour in an agitated pan dryer. As these are normally manufactured up to 8 ft or even 10 ft in diameter, batches of as much as about 100 ft^3 can be dried at a time, eliminating the labour and inconvenience of filling and emptying large numbers of trays. Whereas they are eminently suitable for drying relatively free-flowing materials, they are equally applicable to pastes and even slurries if the design of agitating gear is suitable. Typical pan dryers are illustrated in Fig. 88 and Plate XXXII.

Basically these dryers consist of a shallow cylindrical pan provided with a mechanically driven vertical shaft which carries radial arms supporting scraper blades or ploughs bearing on or just clearing the flat base of the pan. Vertical scraper blades are also frequently fitted which scrape the vertical sides of the pan. The clearance of all scraper blades is usually adjustable. The shaft may be designed for top or bottom drive from a suitable reduction gear and drive. The pan itself may be open-topped or, more usually, covered for atmospheric operation or fitted with a vacuum-tight

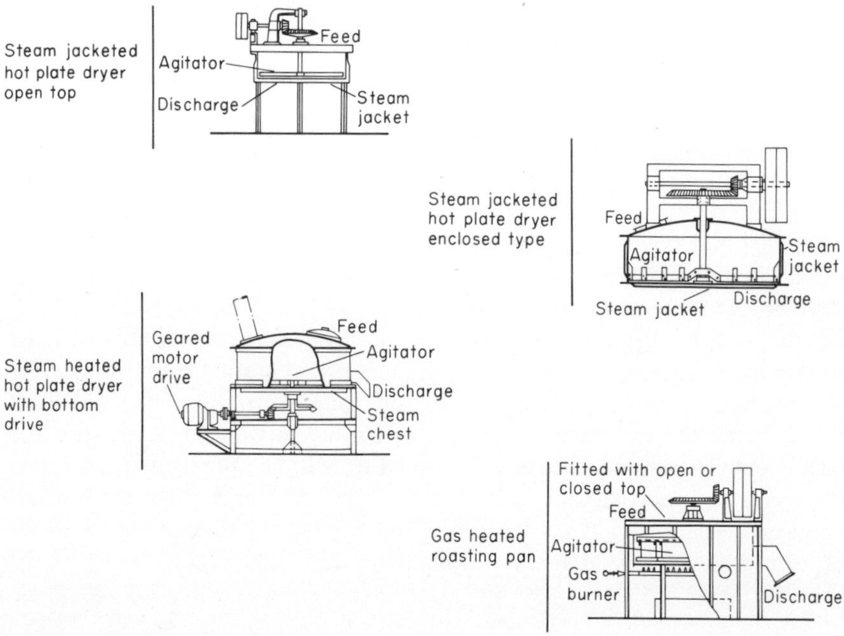

Fig. 88. Agitated pan dryers. *Courtesy A.P.V.-Mitchell (Dryers) Ltd.*

Plate XXXII. Agitated pan dryer. *Courtesy A.P.V.-Mitchell (Dryers) Ltd.*

domed cover for vacuum or reduced pressure operation. Access must be provided in any case for charging and for internal cleaning, and an adequately sized vapour outlet provided. Dry product may be discharged from the base plate or at the base of the side wall, the design of discharge hole cover being such as to operate under atmospheric or vacuum conditions. Where there is any tendency for the mass of charge to rotate with the agitators, it is usual to fit robust metal fingers either from the cover or from the wall. Steam-heated pan dryers are usually constructed with jacketed bottom and sides; with fluid-medium heating or high-pressure steam for high temperatures, heating coils may be cast-in or welded to the walls and base plate. When the dryer is installed in a direct-fired setting it is usual to manufacture the dryer in cast iron of fairly considerable thickness in order to have the maximum uniformity of heat distribution over the area of the pan. In general the materials of construction for the pan, the agitating equipment and the shaft will be selected to withstand the effects of temperature and chemical action during drying.

Unless the feedstock is relatively free-flowing throughout the drying cycle, there is usually danger of product attrition and this may sometimes preclude the use of this type of dryer. However, for most free-flowing and friable materials, the action of suitably designed agitation gear will produce a gentle turning over of the charge and a wave-like movement of the mass, resulting in good material to heated surface and particle to particle contact and consequent uniform heat distribution. When drying under atmospheric pressure it is usual to provide mild air extraction in the form of a fan in the vapour duct in order to maintain the relative humidity of the air over the charge at a low level. Where solvents have to be removed or when drying under a vacuum, a vacuum pump removes released vapours; it is usual to interpose a dust catcher between dryer and pump. When drying slurries and wet pastes clearance between scrapers and heated surfaces must be minimal to prevent a skin of dried material building up on these surfaces, and the whole of the driving gear, together with the shaft and agitating gear, must be designed to contend with the frequently heavy loading imposed on them during the course of drying. This is particularly important when drying high solids content liquids which may go through a peak of considerable power requirement before the semi-dry product becomes reasonably friable as drying proceeds.

Very little published data are available on the performance of agitated pan dryers. Rates of heat transfer vary widely from material to material, being affected by initial and final moisture contents, bulk densities, thermal conductivities and drying characteristics. Assuming heated surfaces are kept well scraped, optimum overall heat transfer coefficients of the order of 30–80 Btu/h/ft^2/deg F are possible in the case of very wet charges, lower figures of from 5–50 Btu/h/ft^2/deg F being more normal for the general run of lower moisture materials. Drying times may range from 3–12 h, largely dependent on the final moisture content required and the ratio of heated surface to volume of charge. Evaporation rates of 3 lb/h/ft^2 are normally the best that can be achieved under atmospheric pressures with fairly high initial moisture content charges; vacuum drying of 50/75% water moist charges may yield higher rates of the order of 5–7·5 lb/h/ft^2. Thermal efficiencies are generally of a low order with direct-firing due to flue gas heat losses, but for steam-jacketed pan dryers they may be of the order of 60–70% under atmospheric pressure conditions and slightly higher under vacuum. All dryer surfaces exposed to the surrounding atmosphere should be adequately lagged.

Table 39. Performance data on agitated pan dryer

	Sodium thiosulphate	*Potassium zeolite*	*Arsenic pentoxide*
Dryer diameter	6 ft 0 in	2 ft 3 in	8 ft 0 in
Dryer depth	2 ft 0 in	1 ft 0 in	2 ft 0 in
Capacity lb product	12 cwt	14 lb	2½ ton/day
Initial moisture, % w/w basis	37	40	35
Final moisture, % w/w basis	0	1	2–3
Method of heating	Steam	Steam	Steam
Atmospheric (a) or vacuum (b)	(b) 26 Hg	(a) 60 lb/in^2 gauge	(b)
Drying temperature: material, °F			
Drying temperature: shell, °F		153°C	
Bulk density product, lb/ft^3			
Drying time h/batch	5	3	8
Material of construction	S.S.	M.S.	S.S.

Courtesy A.P.V.—Mitchell (Dryers) Ltd.

Vacuum Rotary Dryers

These dryers are batch dryers and are supplied in two forms; the more usual has a non-rotating cylindrical jacketed shell housing an internal rotating shaft carrying agitator blades, as illustrated in Fig. 89 and Plate XX, whereas the less usual design shown in Plate XXXIII has a rotating jacketed shell fitted with internal lifting flights.

The dryer shown in Fig. 89 is built in a variety of sizes, depending on production requirements, ranging from 2 ft 6 in diameter by 12 ft 6 in long to 5 ft diameter by 35 ft long. The stationary shell, mounted horizontally, is jacketed, usually in sections, the number of sections depending on the length of dryer shell to ensure uniform temperature throughout its length. Depending on the desired drying temperature hot water, steam or a suitable fluid heating medium is fed to the jacket and, in large dryers, to the

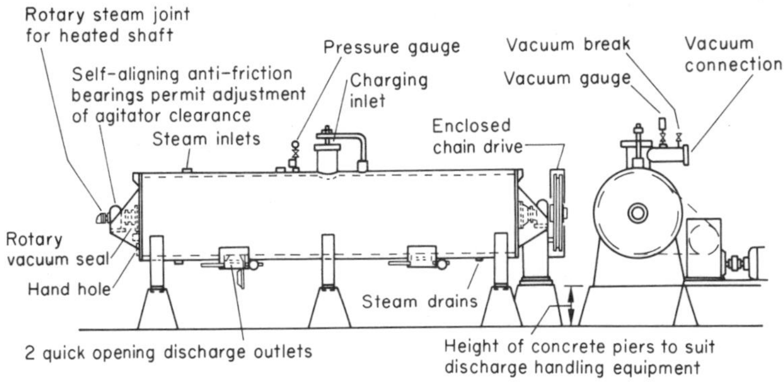

Fig. 89. Vacuum rotary dryer (fixed shell type). *Courtesy Pennsalt Chemicals Corporation, Equipment Division*

Plate XXXIII. Vacuum rotary dryer (rotating shell type). *Courtesy A.P.V.—Mitchell (Dryers) Ltd.*

central shaft and if of suitable design and size to the agitator gear itself. The design of the agitator gear will depend on the nature of the product and may be a discontinuous spiral or a double spiral, or in some instances hollow straight blades. In all cases the clearance between the agitator and the inner shell wall is a minimum and the action of the agitator is designed to give a backward and forward movement of the charge between the dryer centre and its ends. Small dryers have a single central discharge manhole, whereas large dryers have two discharge manholes located usually midway between the dryer centre and its ends. The dryer is charged through one or more manholes mounted above the dryer, a vapour dome of ample dimensions being usually located above the centre of the dryer, leading to a dust trap and thence to a vacuum pump or ejector and condenser when required to remove solvents. Where the central shaft and agitator gear are heated, the heating medium is introduced and removed through an end trunnion. Charging and discharging manholes are of vacuum-tight quick-release design, the latter being usually provided with counter-weighted operating levers.

The dryer shown in Plate XXXIII is not usually built in such large sizes as the fixed-shell design and is usually employed for the drying of smaller charges of relatively free-flowing materials. It is usually provided with a girth ring or tyre carried on rollers and a spur gear and pinion drive, all mounted on a common bedplate. Vacuum is applied through a hollow trunnion which also carries the steam supply and condensate discharge connections for the dryer jacket. Where the dryer is carried by one trunnion and a girth tyre, the other end usually takes the form of a quick-release door for internal cleaning or for emptying the dryer, which is also provided with a manhole, usually centrally through the jacketed shell for charging the dryer. Where the product tends to dust it is not unusual to incorporate a rotating spiral or screw in the trunnion vacuum connection leading to a dust or mud catch-pot. Lifting flights fixed to the internal surface of the shell are customary to promote agitation.

Both types of dryer are used on materials which are heat-sensitive, from which air must be excluded or which contain recoverable solvents. The fixed-shell rotary agitator design can be used for pastes, sludges and free-flowing materials which do not form hard caking on the agitator gear and which will not suffer adversely from attrition caused by the agitators. Where these complications are to be faced, the rotating shell design is preferable although this is today being increasingly displaced by the Double Cone design described later.

The performance of indirect rotary vacuum dryers is very similar to that of agitated vacuum pan dryers and will vary with the nature of the material, its bulk density, weight of charge, initial and final moisture contents and the degree of vacuum which can be applied. The volume and weight of charge will normally run at 45–60% of the dryer volume, less for dense heavy materials, more for light materials. Rotational speeds for the agitator in the first case, the dryer shell in the second are normally low – 3–5 rev/min – although higher speeds improve the heat-transfer rates at the expense of power requirements. Average evaporation rates are of the order of 1 lb/ft^2

of heated surface per hour and steam requirements are about 1·1–1·3 lb/lb of water evaporated. Evaporation rates are higher when drying high initial moisture content charges to a product having a fairly high moisture content. Thus, drying starch from 40/45% moisture to 10% final moisture (w/w basis) may give an evaporation rate of 2 lb/ft^2/h. Due to the constant disturbance of the charge and the presentation of fresh material to the heated surface the evaporation rate is usually four times that experienced in vacuum shelf dryers on the same product. Further, evaporation rates in full-scale plants may be 50% higher than rates determined in small pilot-scale tests, due to the much greater loads of material on the heated surfaces, but drying times in full-scale plants will be longer than those indicated by test units because of the larger charge weight to dryer heated area. The charge temperature will usually remain substantially constant at the boiling point of the moisture under the prevailing pressure during the constant rate period, gradually increasing to that of the jacket temperature as drying proceeds. Drying times are materially affected by the final moisture content desired in the product; it is not unusual that the drying time is doubled for a reduction from say 1 to 0·1%. Similarly, steam consumptions are considerably increased when drying to almost complete dryness, due to the extended drying cycle. Thus, with the drying times to commercially nil moisture content involving drying times of the order of 36 h, the specific steam ratio may rise to 5 lb/lb of moisture evaporated. Drying rates will similarly vary from 2·5 lb/ft^2/h for high initial moisture charges, dried down to moderately low terminal moisture, to 0·1 lb/ft^2/h for materials carried to nil moisture content.

Tumbler Dryers

Horizontal cylindrical rotary vacuum dryers are being largely replaced by tumbler or double-cone vacuum dryers which may, of course, be used as atmospheric batch indirect dryers also. Designed as two opposing jacketed cones on a common jacketed short cylindrical base, this type of dryer has the advantage that with the steep slope of the cones, the dryer, when stationary with the cones in a vertical position, can be rapidly discharged on completion of drying, whereas the rotary agitated vacuum dryer requires continued rotation of the agitation equipment to promote the discharge which, in any event, occupies a very appreciable amount of time. However, in general, agitated horizontal vacuum rotary dryers are more suitable where the charge is not reasonably free-flowing.

A typical Tumbler dryer is illustrated in Plate XXXIV and shown diagrammatically in Fig. 90. The jacketed dryer is provided with supporting trunnions running in suitable bearings, a vacuum connection leading to

Working capacity cu. ft.*	Total volume cu. ft.	H.p.**	A	B	C	D	E	F	G	H
3	4·5	1/2	60"	39"	59"	21"	24"	12"	6"	16"
10	15·4	3/4	72"	51"	80"	30"	36"	14"	8"	18"
25	38·5	3	84"	63"	103"	37"	48"	16"	8"	20"
75	115	7 1/2	128"	81"	117"	50"	72"	16"	10"	20"
100	165	10	140"	93"	135"	60"	78"	18"	12"	22"
150	230	15	157"	106"	140"	66"	90"	18"	12"	22"

* Maximum working capacity normally 65% of total volume but may vary depending on product and conditions
** H.p. shown based on material density of 50 lbs/cu.ft.

Note: All dimensions are approximately interior of vessel designed for vacuum only
Jacket—asme code constructed for 50 p.s.i.

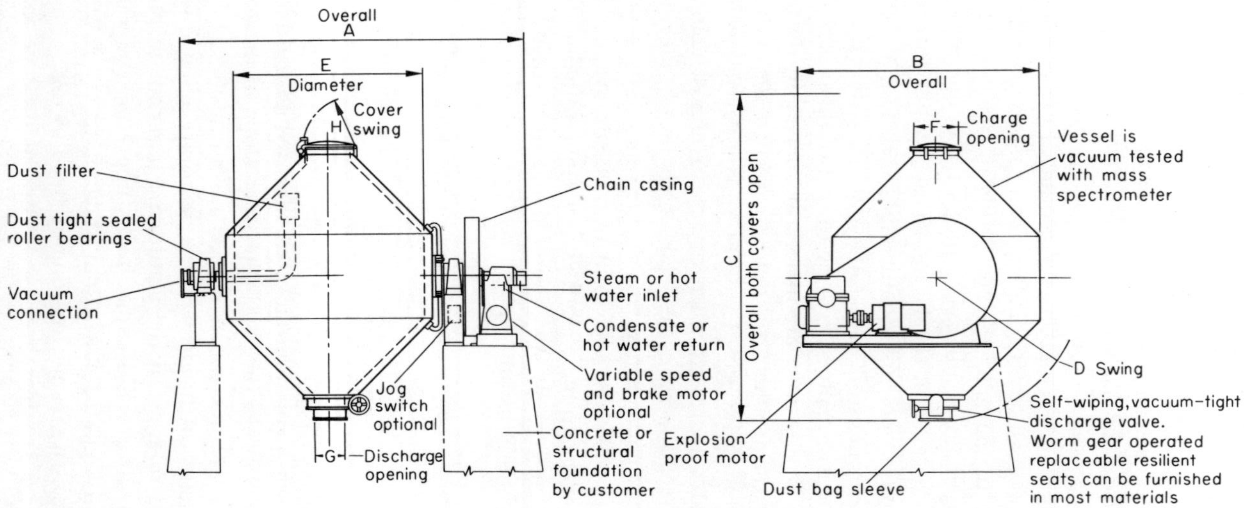

Fig. 90. Tumbler dryer.

Courtesy Pennsalt Chemicals Corporation, Equipment Division

Plate XXXIV. "Conaform" tumbler vacuum dryer.

Courtesy Patterson Division, Banner Industries, Inc.

one trunnion and heating medium (steam, hot water, fluid heat) entry and withdrawal being provided within the other, a special design of rotating valve on the trunnion feeding the heating medium to and withdrawing from the dryer jacket through suitable piping linkages. The vacuum connection is usually continued vertically within the dryer and fitted with a dust filter at its end. One trunnion is mechanically driven through a reduction gear and final chain drive or spur wheel and pinion. A variable speed unit is sometimes fitted. A quick-acting charging cover is fitted to the charging hole of one cone and a vacuum-tight discharge valve, usually of the self-cleaning type, to the other. Tumbler dryers are built with charge capacities of from 1 ft^3 for pilot-scale work up to several hundred cubic feet for commercial operation, the charge usually occupying from 60–70% of the dryer internal volume. Rotational speeds will range from 12 rev/min for small units to 3 or 4 rev/min for large commercial installations.

Laboratory tests are essential for determining the drying time and size of an industrial unit. Even so, scale-up factors are a matter of experience. The controlling factors, other than the drying characteristics of the material, the final moisture requirement and the degree of vacuum employed, are heat-transfer area of the dryer, batch size and installed equipment size. With very large batches, the material particles deep in the batch will be under considerably higher pressures than those at the charge surface at any one moment in the tumbling effect but will be varying all the time during the drying cycle. This effect is, however, found in practice to be of considerably less importance than the available heated surface. Whereas the drying time will be inversely proportional to the temperature difference between the dryer heated surfaces and the charge, it is found to be directly proportional to the ratio of cubic feet of charge to the total heat-transfer area. The surface area of the double-cone dryer increases as the 2/3 power of its capacity whereas the particle area is proportional to the size of charge. Increase in dryer size results in a lower increase in heated surface area compared with charge particle area, so that drying times in large commercial units will be materially longer than those obtained under pilot-scale conditions. Fischer[20] states that, other factors being equal, drying times are proportional to batch size divided by jacketed shell area or to the cube root of the tumbler dryer volume. The approximate relationship between small-scale test results and those to be anticipated in full-scale plant are shown in Fig. 91.

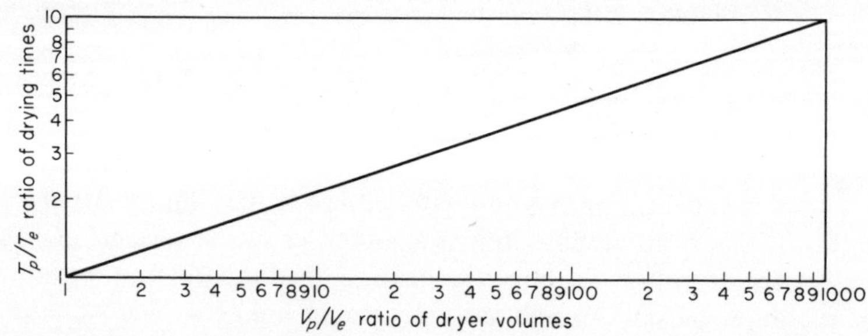

Fig. 91. Tumbler dryer scale-up from laboratory tests.

(after John J. Fischer. Ind. Eng. Chem. 55. 2 Feb. 1963)

Tumbler dryers are today widely used in the pharmaceutical, fine chemical, plastics, pigment and dyestuffs industries for batch-drying operations. Drying performance will be substantially similar to that experienced with rotary vacuum dryers but with shorter turn-round times between charges.

Table 40. Conical vacuum pilot-scale dryer performance

	Tungsten carbide	*Polyester resin*	*Penicillin*	*Hydro-quinone*	*Prussian blue pigment*
Volatile ingredient	Naphtha	Water	Acetone	Water	Water
Physical nature of charge	Heavy slurry	Pellets	Powder	Powder	Filter-cake
Dryer diam, ft	2	2	2	2	2
Dryer capacity, ft^3	2·5	2·5	2·5	2·5	2·5
Method of heating	Hot water	Steam	Hot water	Hot water	Steam
Heating medium temperature, °F	180	240	140	150	225
Vacuum, mm Hg abs	40 to 84	12 to 18	40	50 to 100	40 to 110
Initial volatile content, % w/w basis	18·0	0·34	27·9	5·0	83
Final volatile content, % w/w basis	nil	0·01	nil	0·25	4·8
Weight of charge, lb	640	130	55	61	142·5
Bulk density of charge, lb/ft^3	256	51·5	21·5	26·5	58·5
Drying time, min	155	215	90	50	480

Courtesy Patterson Division, Banner Industries Inc.

Continuous Dryers

Screw-conveyor Dryers

Although not widely used in industry, this type of dryer may have an important part to play particularly when used, as its name implies, as a material conveyor from point to point and a medium degree of drying has to be accomplished at the same time. In its most usual form it consists of a jacketed or heated trough, either open-topped or closed in which a paddle or screw conveys the product in one or more stages from the feed point to the discharge point. Such dryers are usually applied to free-flowing materials. However, when the jacketed part is in a tubular form, or when the screw is itself hollow and fed with a heating medium, non-free-flowing but friable wet materials can be effectively dried and conveyed. When totally enclosed it is usual to apply a mild suction in order to remove released vapours. The nature and temperature of the heating medium will depend upon the heat sensitivity of the feedstock. In multi-section dryers of this type each section can be mounted one above the other or consecutively in line depending upon what direction and elevation at the discharge point is required. Material may be conveyed considerable distances in this type of dryer as, for instance, in the fishmeal industry. When provided with special designs of feed and discharge valves they can be used under vacuum for particularly heat-sensitive materials. They can similarly be used for drying materials wet with solvent which require to be recovered.

The Holo-flite Dryer

Greatly improved performance is obtainable where the means of conveying is in the form of one or more hollow helical screws to which, together with the jacket, the heating medium is fed. Fig. 92 shows a sectional view of such an arrangement. The hollow, heated screws are arranged to intermesh, thereby producing the maximum possible relative movement

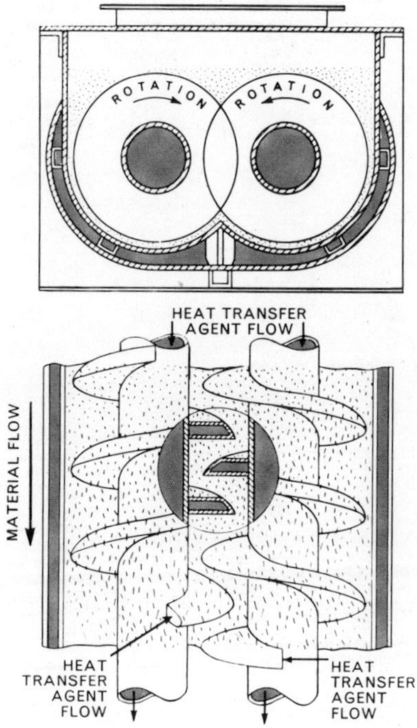

Fig. 92. The Holo-flite dryer. *Courtesy W. C. Holmes & Co. Ltd.*

between heated surfaces and feedstock during the drying and conveying operation. The slow conveying, interfolding and dispersing action of the intermeshed screws is claimed to result in a minimum of breakdown of the dried material. Cooling may be combined with drying by feeding cold water to the last stage of the dryer installation. Typical performance data on the standard range of single-stage Holo-flite dryers are given in Table 41.

Where multi-stage screw conveyors are used on material which tends to compact at the end of the early stages, care must be taken that the discharge aperture from one stage to the next is of ample dimensions for the material to fall through and not build-up, thereby imposing an excessive load on to the conveyor and even causing mechanical damage.

Table 41. Holo-flite dryer performance data

	Material treated: free-flowing solids or slurries	
	Minimum	*Maximum*
Material temperature, °F	20	2,000
Material flow, lb/h	250	90,000
Sweep gas flow, Std ft^3/min	4	20
Operating pressure, lb/in^2 gauge	0	5
Material inlet moisture, % w/w	0	70
Material outlet moisture, %w/w	0	30
Process heat load, Btu/ton	9,000	50,000
Available heat-transfer area, ft^2	22	450
Power requirement, hp/ton/h	0·5	3·3
Peripheral speed of rotating screws, ft/s	0·03	1·0
Dimensions: Length, ft	12·0	26·0
Width, ft	1·0	5·0
Height, ft	1·0	3·0

Courtesy W. C. Holmes & Co. Ltd.

Heat-transfer rates will be somewhat similar to those experienced with other forms of indirect-heated dryers, but will depend upon the speed of rotation of the paddles or screws, increasing with conveyor speeds from a low figure of 2–3 Btu/ft^2/h/deg F to 25–30 Btu/ft^2/h/deg F for high conveyor speeds of the order of 150–200 rev/min.

The Rotadisc Dryer

This is an interesting agitated continuous indirect dryer which was first developed for the fishmeal industry where a sticky press-cake had to be handled. It has since found important application for most meat residues, spent grains in breweries and distilleries and cotton-seed protein. It is equally suitable for free-flowing materials and, although normally operated at normal atmospheric pressure can, with suitable feedstock and dried product valves, be operated under some degree of vacuum. Essentially this dryer, which is illustrated in Fig. 93, consists of a stationary cylindrical jacketed shell in which a hollow rotating shaft carrying a large number of smooth-surfaced hollow discs is fitted. Up to 85% of the total heating surface of the dryer is concentrated in the shaft and rotating discs so that a very large heating surface is confined in a unit of compact size.

Each disc carries at its circumference a number of adjustable agitator blades which, with the rotation of the discs, propels the material from feed point to discharge point. The highly polished disc surfaces are kept clean by stationary scraper blades carried from the inner shell of the dryer and projecting between each pair of discs. Access and cleaning man-

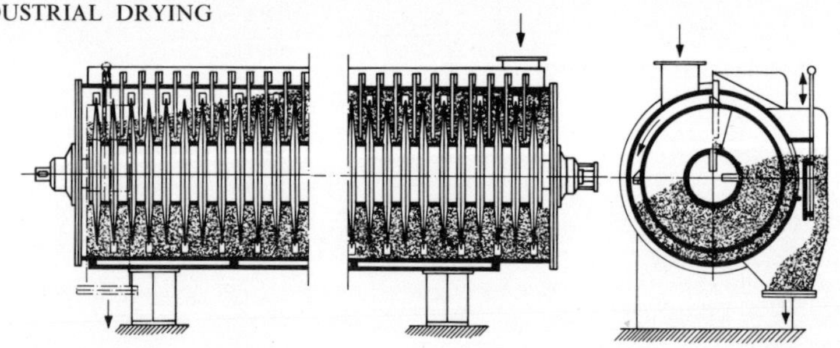

Fig. 93. The Rotadisc dryer. *Courtesy Stord-Bartz Industri S.A.S.*

holes are provided on the upper surface of the shell through which the progress of the feedstock through the dryer may be viewed. Designed for normal maximum steam pressure of 85 lb/in^2 gauge the evaporation rates average about 1·6–2 lb/ft^2/h with specific steam ratios of 1·2–1·4 lb steam/lb water evaporated. Standard commercial units have from 200 to 3,000 ft^2 of heated surface.

Direct Agitated Dryers

An interesting agitated continuous dryer, in which the heat transfer and drying is by means of direct contact between a gaseous medium such as combustion gases and the feedstock, is the Hazemag Rapid dryer illustrated diagrammatically in Fig. 94. The dryer body itself is stationary

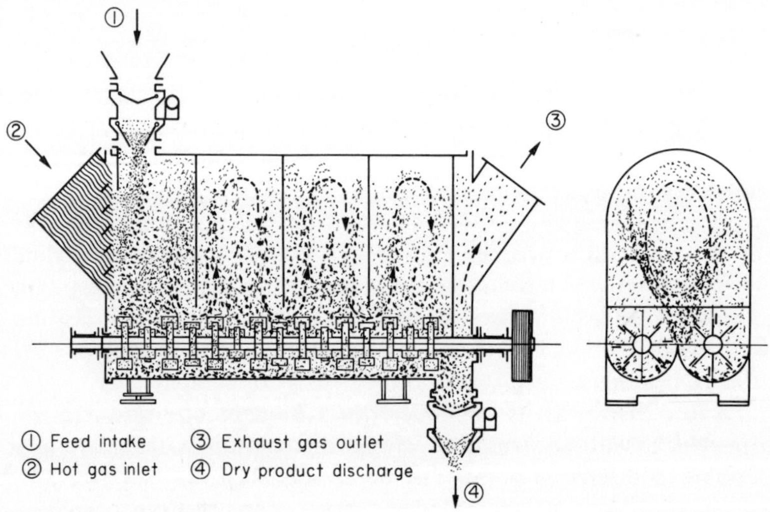

Fig. 94. The Hazemag Rapid dryer. *Courtesy W. C. Holmes & Co. Ltd.*

and is a steel shell provided with a double pendulum self-sealing feed connection and dry product discharge together with hot gas and exhaust gas connections. Material is propelled through the dryer by either one or a pair of rotating shafts carrying paddles, depending on the size of the installation. The paddles, which are easily replaceable, are made of wear-resisting material, and are rotated at a relatively high speed, thereby throwing the feedstock into the gas stream and ensuring intimate contact therewith. This dryer is widely used for clays, filter cakes and minerals and typical data relating to standard-sized units are given in Table 42. This dryer is normally operated on the co-current principle and a typical heat balance of an industrial installation is given in Fig. 95.

Table 42. Hazemag Rapid dryer performance data

Material treated: Clays, filter cakes, minerals	*Minimum*	*Maximum*
Gas temperature, °F	840	1,470
Material flow, lb/h	1,000	300,000
Sweep gas flow, Std ft^3/min	350	14,000
Normal operating pressure	slight vacuum	slight vacuum
Material inlet moisture, % w/w	normally 5	40 or slurry
Material outlet moisture	as required	as required
Material size range at feed, in	dust	4
Efficiency, %	80	90
Process heat load, Btu/ton	130,000	1,200,000
Pressure drop through dryer, in w.g.	2	6
Power requirements, hp consumed/ton/h	1·0	2·0
Peripheral speed of paddles, ft/s	18	24
Dust burden in exit gases, grains/ft^3	5	2,000
Dimensions: Length, ft	8·0	25·0
Width, ft	2·7	13·0
Height, ft	2·7	13·0
Evaporation rate, lb/h	100	25,000

Courtesy W. C. Holmes & Co. Ltd.

Vacuum Pan Type Dryers

This dryer, illustrated in Fig. 96, is used for friable, free-flowing products which have to be dried under vacuum, have recoverable solvents and are continuously produced, thus justifying a continuous dryer. The dryer operates in the same manner as a batch-agitated pan dryer except that the feedstock is fed on to a stationary jacketed plate or platen on which the material is agitated and moved round to a discharge point from which it falls to similar heated platens arranged tierwise, each having its own scraper-agitator driven from a central vertical shaft. Heating medium is fed to each tier of platens by suitably arranged internal heaters and condensate or spent fluid heating medium is removed in the usual manner.

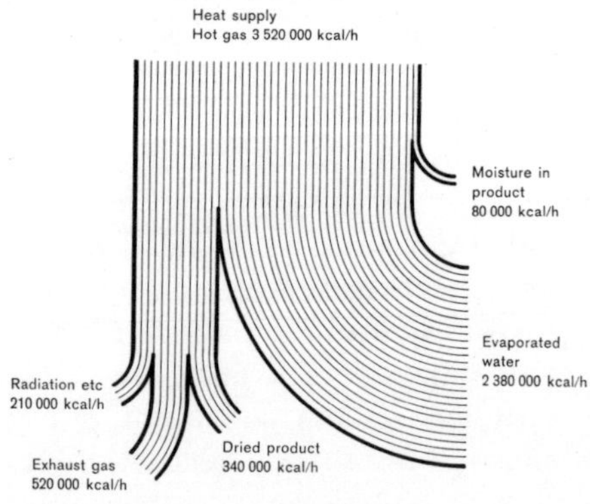

Fig. 95. Typical heat balance of Hazemag Rapid dryer. *Courtesy W. C. Holmes & Co. Ltd.*

Specially designed pneumatically operated lock-valves are used to charge and discharge the dryer in a semi-continuous manner, the rate of feed and discharge being automatically controlled to suit the throughput from a control console. These dryers are available in four standard sizes having heated platen surfaces ranging from approximately 4–950 ft^2 in contact with the feedstock and requiring from 1·1–7·5 kW motive power. Typical performances of this design of dryers are given in Table 43.

Table 43. Performance of continuous vacuum pan dryers

	Sodium hydrosulphite	*Maneb*	*Melamine*	*Activated carbon*
Dryer diameter (pans), metres	2	2	2	2
Number of pans	5	17	11	17
Area (approx) m^2	12·4	42·8	27·6	42·8
Dry product, lb/h	1,100	660	1,870	440
Initial moisture, % w/w	4	23	11	62
Final moisture, % w/w	0·1	0·5	0·03	3
Heating	Hot water	Steam 1·3 atm	Steam 2·5 atm	Steam 2·5 atm
Pan temperature, °C	98	105	125	125
Evaporation rate, lb/ft^2/h	0·325	0·325	0·79	0·78
Drying time, min	15	170	12	30

Courtesy Krauss-Maffei-Imperial G.m.b.H.

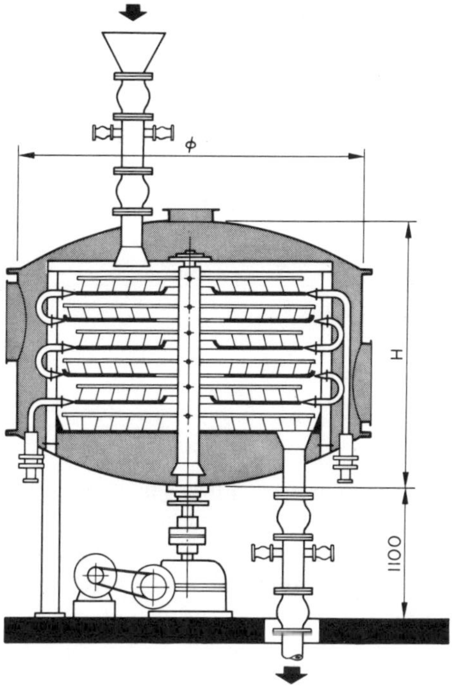

Model	Pan surface m^2	φ mm	H mm	Power consump-tion kW	Model	Pan surface m^2	φ mm	H mm	Power consump-tion kW
12/5	3·8	1900	1570	1·1	20/23	57·9	2500	4660	5·5
20/5	12·4	2500	1780	2·2	20/29	73·1	2500	5620	7·0
20/11	27·6	2500	2740	3·0	20/35	88·2	2500	6580	7·5
20/17	42·8	2500	3700	4·0	30/...				

Note: $1 m^2 = 10·76$ sq. ft.
1mm = 0·03937 inch
1kW = 1·341 h.p.

Fig. 96. Continuous vacuum pan dryer. *Courtesy Krauss-Maffei-Imperial G.m.b.H.*

CHAPTER XIII

Miscellaneous Drying Systems Infra-red, Freeze and Dielectric Dryers

Infra-red Dryers

Whereas the direct-heated and indirect-heated dryers described in previous chapters rely essentially for heat-transfer and drying effects on convected and conducted heat, respectively, with radiant heat playing an incidental – if occasionally important – role, infra-red dryers are designed to utilise radiant heat to the maximum extent. Heat transfer by convection and conduction in drying equipment is normally at relatively low levels of temperature and is a function of the temperature difference between the heating medium and the material processed at these levels. As has been shown, even at these levels radiation of heat can assume important dimensions; at high heat source temperature, heat transmission by radiation is a function of the difference between the fourth power of the heat source and irradiated body absolute temperatures. At heat source temperatures, ranging from dull red to incandescence, heat-transmission rates will obviously be very considerable. Heat source temperatures in infra-red dryers are generated by gas-heated surfaces and tubular encased resistance wire elements at the lower levels and by incandescent tungsten filament lamps or tubes at the highest levels. Black panel and incandescent gas-heated refractory elements may also be used. Infra-red radiant heat energy is electromagnetic radiation of wavelengths longer than visible light energy, but shorter than radio waves, and which lie between 0·7 and 1,000 microns.

The visible spectrum terminates at 0·7 micron wavelength. So-called "near" infra-red sources are defined as sources having peaks ranging from 0·7 microns to around 2 microns. Heat sources which have peak radiations longer than 2 microns but below 7 or 8 microns are referred to as "far" infra-red sources. Sources having thermal radiations longer than 8 microns have low heating properties and are therefore of little interest in heating or drying applications. In general it may be said that "near" infra-red energy is generated at heat source temperatures above 2,500°F, below which energy radiation will lie in the "far" infra-red region. As the heat source temperature increases, the peak of the radiant energy curve moves towards the visible spectrum, the wavelengths becoming shorter, so that incandescent infra-red heat sources will to some extent overlap the visible spectrum and the heat source will have some visible light output. As the heat source temperature increases, the amount of energy radiated increases very considerably and follows the Stefan-Boltzmann Law, which states that the radiant output is proportional to the fourth power of its absolute temperature.

Electric Power Energised Sources

Incandescent tungsten filaments in globular lamps and translucent quartz tubes generally operate at colour temperatures around 4,000°F as compared with about 4,700°F for lamps for illuminating purposes. Lamps are generally supplied up to 500 or 750 wattages whereas linear type transclucent quartz tubes may have power ratings of from 100 up to as high as 350 or 360 W/linear in and offer the highest heat concentration available. The infra-red energy generated is at wavelengths which pass through quartz and glass with only about 15% loss of input energy as radiant heat, the remainder being convection and conduction heat losses. Lower rating quartz tube elements at about 50 W/linear in power are also used and these may have a radiant efficiency of only 50–65%, the operating temperature being around 2,000°F at which temperature the heat radiation is at longer wavelengths of the order of 4 microns, resulting in greater absorption by the quartz envelope and re-radiation of heat energy at even longer wavelengths. Metal sheath elements which operate at the far end of the spectrum at temperatures of the order of 1,000 1,800°F consist of a Nichrome or similar resistor wire element surrounded by a magnesium oxide insulator and a tubular metal enclosure to which heat is conducted and from which heat is radiated as radiant heat to the extent of from 45–60% of the total energy input, the remainder being dissipated by convection and conduction to surroundings. They are, however, extremely rugged and generally have a long useful life which makes them attractive in some cases as compared with glass lamps. Quartz tube elements are less liable to fracture than glass lamps and, where fragmented glass becomes a process hazard—as in the food industry—and high intensity radiant heat is desirable, quartz tubes are being increasingly adopted. Gas-heated infra-red heat sources available today operate at the far end of the spectrum. Infrared sources of heat of very high thermal outputs are today practical, with long useful life. Translucent quartz tube elements with incandescent tungsten filaments can be spaced at close centres; 100 W/linear in elements spaced at $\frac{9}{16}$ in centres will produce a heat density of 25 kW/ft^2 of heater surface, equivalent to a radiant output of 75,000 Btu/h/ft^2 assuming a radiant heat output efficiency of 86–88 %.[21] Gas-heated infra-red equipment is also available with radiant heat outputs of 20,000–30,000 Btu/h/ft^2. At design operating temperatures of 4,000°F the useful life of a quartz lamp or tube is of the order of 5,000–6,000 hours.

The radiant heat energy emitted from the heat source requires to be directed on to the work which has to receive this heat and for this purpose reflectors are used. In the case of globular lamps the reflector is often built into the lamp itself but more usually in infra-red dryer designs the reflector, generally of parabolic shape, is separately constructed in multiple banks to suit the contour of the heat-receiving work. Reflectors of 100 % efficiency are not attainable in practice but reflectors in specular gold deposited electrolytically over coatings of bright nickel and copper on a base form of steel have an efficiency of 97–98 %. Silver and copper, although high in infra-red reflectivity, oxidise too easily; aluminium which has been processed to resist oxidation is more suitable and is frequently used, having a reflectivity to infra-red radiations of about 90 %. Chromium-plated and stainless steel reflectors although otherwise suitable are inferior infra-red reflectors. Aluminium painted steel reflecting surfaces have an infra-red radiant heat reflecting efficiency as low as 50 %. The higher the reflectivity of the reflectors employed in constructing the oven the less will be the dissipation of heat as convected and conducted heat – some of which can be usefully recovered and applied to the drying operation – but the higher initial cost of the specular gold reflectors may not always be justified as compared with reflectors in anodised or similarly processed aluminium. Reflectors – whether for globular lamps or for linear heat source tubes – are generally mass produced in modular units or sections which permit building walls, enclosures, canopies or tunnels with relative ease and in such a form as to suit the geometry of the processed work and to secure maximum re-reflection of stray radiations on to the work. Fig. 97 shows the theoretical radiant heat utilisation for products of varying emissivities with change in reflectivity of source reflectors and surrounding wall reflectivity.[23]

The capacity for absorbing infra-red radiant heat is affected by the colour of the irradiated surface, that of black (lamp black) being the highest and white the lowest. The radiant heat absorptivity of a body is the fraction of incident radiant heat which is absorbed and is at its maximum, namely unity, for a perfect black body having no reflectivity. Kirchhoff's Law states that at thermal equilibrium the ratio of emissive power of a surface to its absorptivity is the same for all bodies. The ratio

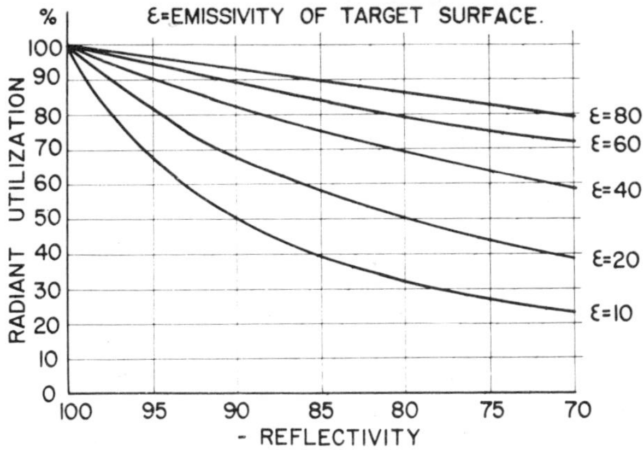

Fig. 97. Theoretical radiant heat utilisation for products of varying emissivity with change in source reflector reflectivity. *(after Robt. W. Jeffrey. Adhesive Age. Aug. 1968 pp. 29–34)*

of the emissive power of a material surface to that of a black body is called the emissivity of that surface and at the same temperature is equal to its absorptivity. The darker the colour of the irradiated surface in general, the greater its absorptivity and the less its non-absorption or reflection of the radiant heat. Considering heat sources alone, the low-brightness far infra-red types, such as quartz tubes, metal rods and heated black panels, will heat bodies of different colours at more nearly the same rate than will the high brightness or near infra-red types such as quartz lamps and glass lamps. However, the heat source is only a part of a complete oven and the greater colour sensitivity of near infra-red heat sources can be compensated for by the use of reflectors of maximum reflectivity and correct disposition of reflecting surfaces, so that the heat rays which miss the irradiated object on their first reflection are usefully absorbed on subsequent reflections. The result is maximum application of radiant energy, and reduction in heating rates due to the lower absorptivity effects of different colours becomes less important. The degree of penetration of infra-red radiant heat energy depends on the absorptivity characteristics of the irradiated body as well as on the wavelength of the radiation. Thus the "red" heat generated by Nichrome wire heat source elements at about 1,900°F has less penetration than the near infra-red radiations from tungsten heat sources, resulting in a greater surface heating effect, which may make it more desirable for skin heating for surface drying. Quartz lamps and tubes employing tungsten filament heat sources reach peak energy outputs in the 1·0–1·5 micron wave band where penetrating power is high. Energy peaks on an equal input basis are illustrated in Fig. 98 for heat sources ranging from

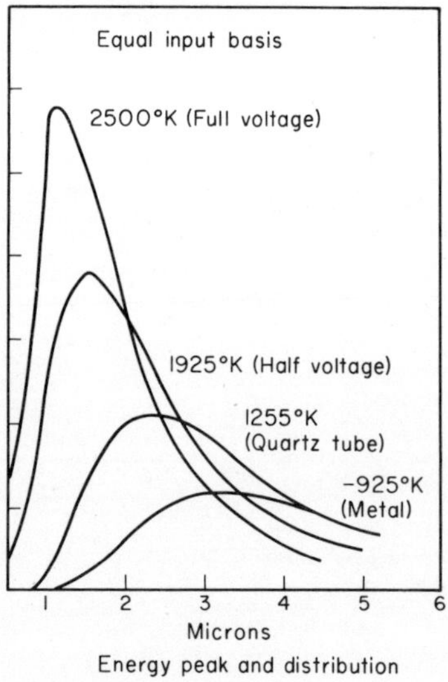

Fig. 98. Radiant heat energy peaks and distribution. *Courtesy Fostoria-Fannon Inc.*

quartz tungsten filament units at about 4,030°F (2,500°K) through 3,186°F (1,925°K) for the same elements at half-power rating, 1,800°F (1,255°K) for low rating quartz tube elements and 1,200°F (925°K) for red heat metal tube elements together with the areas in the wave bands at which these peaks occur.

Many of the applications of infra-red heating in drying operations call for rapid build-up to peak temperatures on the one hand and equally rapid cooling down at a certain or final stage of drying on de-energising. Large automobile paint shop installations require rapid maximum temperature response on start-up, whereas textiles could be ruined and a fire might result if the cooling down of the heat source were not almost instantaneous on de-energising in the event of a mechanical stoppage. Tungsten filament sources have a most rapid response at switch-on and switch-off and Fig. 99 gives a comparative set of curves for tungsten filament and Nichrome wire elements.

The characteristics of the more usual type of infra-red heat sources are compared in Table 44.

Oven Design

Infra-red dryers are usually, but not always, applied to materials and components supported on or from a conveyor system, the dryers taking the form of a tunnel designed to conform as closely as possible to the

Table 44. Characteristics of commercially used infra-red heat sources

	Tungsten filament wire		*Nichrome spiral winding*		*Low temperature panel heater*	
	Glass bulb	*T3 quartz lamp*	*Quartz tube*	*Metal sheath*	*Buried nichrome*	*Metallic salt*
Source temperature						
Normal max.	4,000°F	4,000°F	1,600°F	1,200°F	600–800°F	
Usual range	3,000 to 4,000°F	3,000 to 4,000°F	1,800 to 1,400°F	1,400 to 1,000°F	1,100–400°F	
Brightness	Bright white heat	Bright white heat	Cherry red	Dull red	No visible light	
Usual size	G-30 lamp	$\frac{3}{8}$ in dia. tube	$\frac{3}{8}$ in or $\frac{5}{8}$ in dia. tube	$\frac{3}{8}$ in or $\frac{5}{8}$ in dia. tube	Flat panels – various	
Wavelength at energy peak						
Normal max.	1·15 microns	1·15 microns	2·6 microns	3·1 microns	Around 4–5 microns	
Usual range	1·5 to 1·15 microns	1·5 to 1·15 microns	2·6 to 2·8 microns	2·8 to 3·6 microns	3·2 to 6 microns	
Relative energy distribution						
Normal maximum:						
Radiation	80%	80%	55%	50%*	40–30%*	
Convection and conduction	20%	14%	45%	50%*	60–70%*	
Usual range:						
Radiation	60–80%	72–86%	55–45%	53–45%*	50–20%*	
Convection and conduction	35–20%	28–14%	45–55%	47–55%*	50–80%*	
Degree of heat penetration	Depth of penetration varies with the characteristics of the product. As a general rule, energy of shorter wavelengths penetrates deeper than energy of longer wavelengths.					
Relative response						
to heatup	seconds	seconds	minutes	minutes	scores of minutes	
to cooldown	seconds	seconds	seconds	minutes	scores of minutes	
Colour sensitivity	Bodies of different colours can be heated at more nearly the same rate by infra-red radiation with long wavelengths than they can by short wavelength infra-red radiation.					
Ruggedness						
Mechanical shock	poor	good	good	excellent	varies with panel design – could be quite good	
Thermal shock	poor	excellent	excellent	excellent		

Courtesy Fostoria-Fannon Inc.

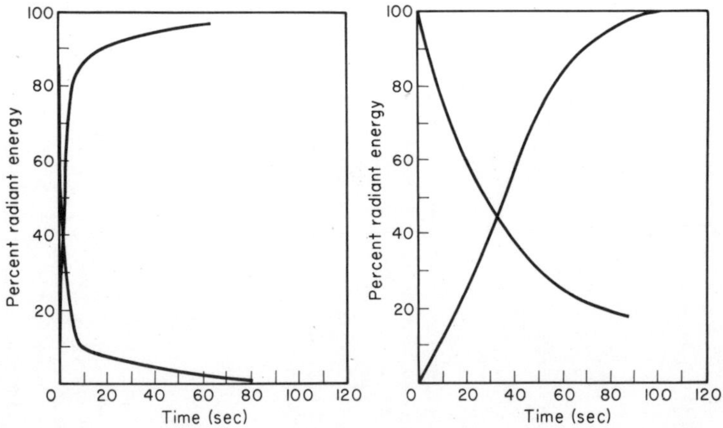

(a). Heat-up and cool-down rate tungsten lamp.

(b) Heat-up and cool-down rate nichrome

Fig. 99. Heat-up and cool-down rates for tungsten filament and nichrome wire heat sources.
Courtesy Fostoria-Fannon Inc.

contour of the work processed. They may take the form of flat panels of radiants on either side of the suspended components, or an inverted U form, or may be flat horizontal banks of elements superimposed above the work and if necessary radiating upwards from below. Where there is a danger of moisture dripping on to the elements these will usually be of quartz as glass lamps would suffer fracture. In all cases the dryer is usually designed for accessibility to the heat source and reflectors for purposes of cleaning, replacement and general maintenance. In order to cut down convection losses, the dryer enclosure should be adequately insulated. Where necessary, especially when moisture is in the form of organic solvents as in paint drying and baking, the installation will be provided with an exhausting system, and where there are considerable re-entrant angles to the shape of the work a measure of internal forced convection may be employed to obtain uniformity of material temperature. As far as possible, the installation must be designed so that all areas of the work have a view of the heat source and reflector surfaces. Typical infra-red dryer types are shown diagrammatically in Fig. 100 whilst Plates XXXV and XXXVI show two typical industrial installations.

Glass lamp emitters are generally used for relatively low temperature drying—up to about 400°F—and provision is made for either natural or forced convection air cooling of the lamp sockets and wiring. Higher temperature drying—up to 1,200°F and above—requires forced air or even liquid cooling of connections and reflectors.

As heat transmission is essentially by radiation, the irradiated material attains temperatures, particularly on the surface, considerably above the wet-bulb temperature of the atmosphere in contact with it and this form of

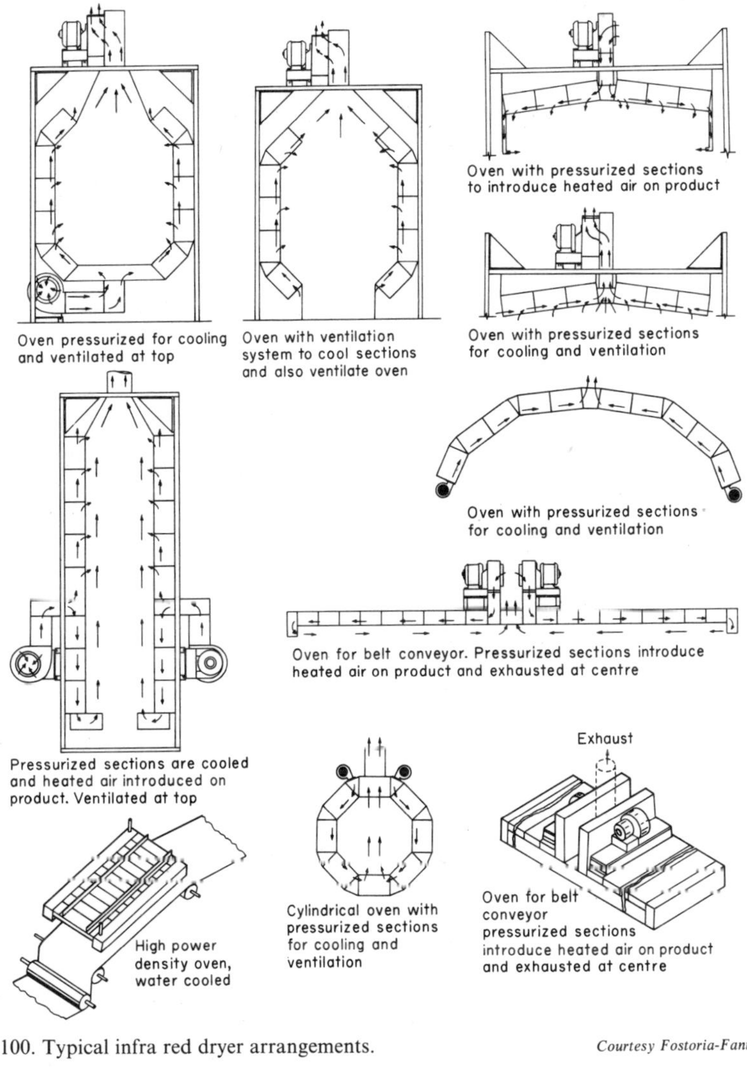

Fig. 100. Typical infra red dryer arrangements. *Courtesy Fostoria-Fannon Inc.*

drying is not suitable where internal moisture diffusion is controlling or where there is substantial shrinkage on drying. It is most suitable for surface drying and thin film drying although it may be applied to small particles and granules submitted to vibration so as to expose new surfaces constantly to the radiant heat. Drying will depend upon the intensity of radiation (flux density); the absorptivity of the material; its heat capacity, and the proportion of heat energy usefully imparted to it from the heat

Plate XXXV. Infra-red dryer for vehicle surface coatings. *Courtesy Fostoria-Fannon Inc.*

source. Tiller and Garber,[23] who studied the heating of thin metal sheets by infra-red radiation, found that the maximum temperature attained by the sheet could be represented by

$$t_a + \frac{3{\cdot}41\,EI}{Yh_c}$$

where t_a = the temperature of the surrounding atmosphere, °F
E = the absorptivity of the material
I = the intensity of radiant heat on the surface, watts/ft^2
Y = the ratio of area of sheet subject to convection to area receiving direct radiation
h_c = the convection heat transfer coefficient, Btu/h/ft^2/deg F.

A close approximation for calculating the heat source wattage for an infra-red drying operation is given by

$$Fc_p \frac{(t_2 - t_1) + W\lambda}{3413\,XE}$$

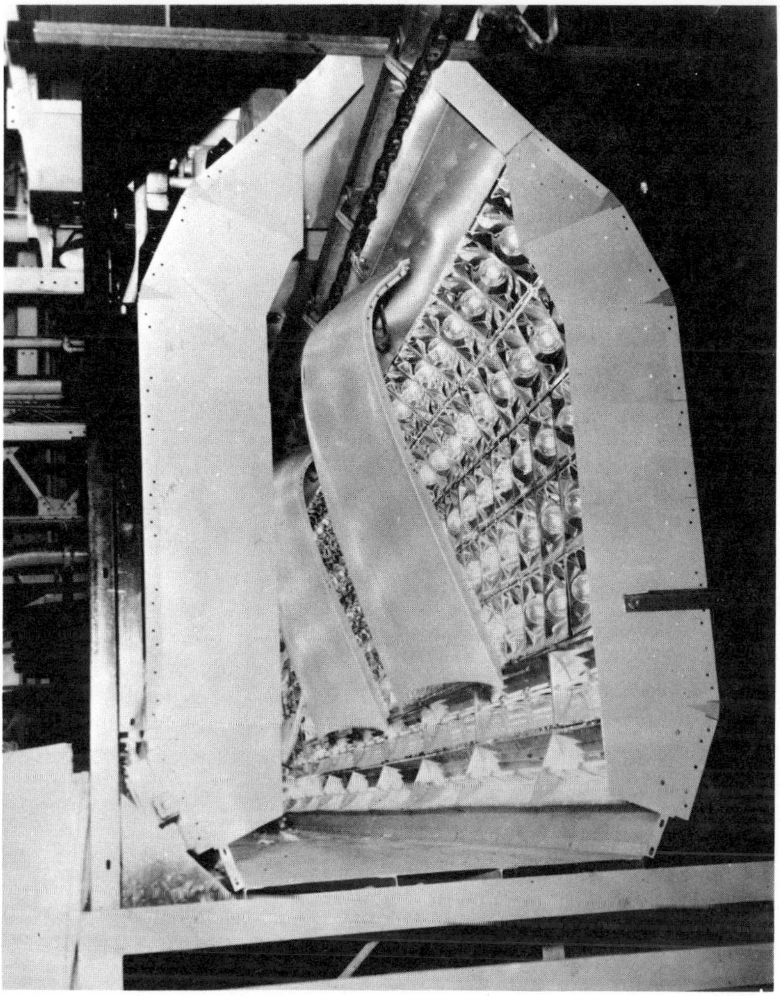

Plate XXXVI. Continuous infra-red oven for automobile parts. *Courtesy Fostoria—Fannon Inc*

where F = the feed rate of moist stock, lb/h
c_p = its specific heat at constant pressure, Btu/lb/deg F
t_1 = the work initial temperature, °F
t_2 = the work final temperature, °F
W = the moisture evaporated, lb/h
E = is the absorptivity of the work surface
X = heat source/reflector overall efficiency times the space factor

The combined heat source-reflector efficiency factor will be highest for lamp/gold reflector and lowest for metal tube elements, ranging from 0·8 for the former to 0·45 for the latter. For a well-designed oven the space factor can be taken to be 0·85 to 0·9. The absorptivity of the work will vary widely depending on its nature and degree of polish, but for most purposes will range from 0·15 to 0·9 for oxidised near-black iron and steel. Typical absorption factors for some common materials are given in Table 45.

Table 45. Radiant heat absorption factors for common surfaces

Material surface	*Temperature range °F*	*Absorption factor*
Polished aluminium	60–70	0·04
Oxidised aluminium	400–1,000	0·1–0·19
Polished copper	60–70	0·03
Iron, steel		
Polished	1,000	0·15
Machined cast iron	1,500	0·53
Oxidised (rust)	70	0·6
Black sheet and ingot	1,500–2,000	0·9
Painted in colours	—	0·6–0·75
Painted in aluminium	200	0·3–0·5
Pottery (glazed)	—	0·85–0·9

Data from which an infra-red drying oven can be designed must be obtained by laboratory testing to ascertain the radiant energy concentration required from a particular heat source. Greatly varying thicknesses of section of the work may necessitate the slowing down of the heating rate to permit of temperature equalisation as, if the results are to be used for the design of a continuous dryer, it may be difficult to arrange areas of higher flux density to compensate for configuration of radiants and reflectors about the contour of the work and their distance from the work. Radiant heat concentrations over the drying cycle may have to be varied to correspond with heating-up, holding and cooling-down zones in the final commercial installation and even in areas of the cross-section.

Temperatures attainable with heat source radiation intensities obtainable today are indicated in Fig. 101 where the "black body" temperature of the product is plotted against the radiation intensity in kW/ft^2.[21]

Infra-red heating for purposes of drying has many advantages where found suitable. High radiant heat concentrations are derived from sources occupying very little space. Heat source temperatures are attained extremely quickly and there can be little or no retention or residual heat on shut-down. Complications arising from products of combustion – except with gas-heated sources – are absent and atmospheres surrounding the work are clean. An absolute minimum of floor space and height is required, equipment lends itself to easy conveyorising, maintenance and cleaning is

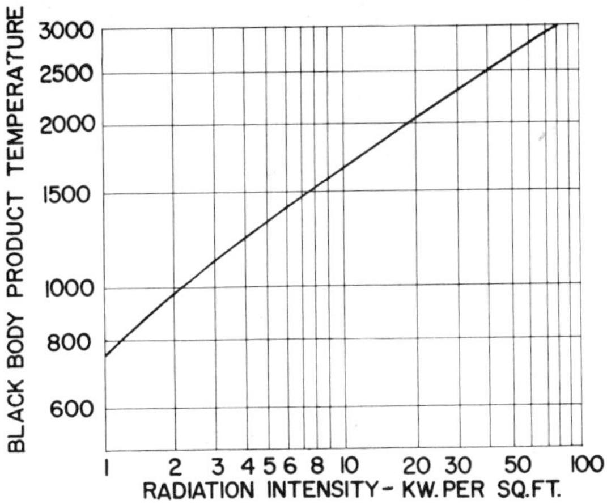

Fig. 101. Product temperature—power density relationship.

(after Robt. W. Jeffrey. Adhesive Age. Aug. 1968. pp. 29–34)

minimal and a correctly designed and installed plant is amenable to accurate and instant response to automatic energy input control. For these reasons infra-red systems are widely used for heating and baking operations and have considerable applications for drying, particularly for removal of surface moisture, surface coating solvent removal and the drying of such products as pharmaceutical granules and food products of relatively low material thicknesses.

Control

The control of infra-red drying and heating installations has until recent years been largely limited to the operation of on–off switches but great advances have been made latterly with the more widespread use of such plants in industry, particularly on continuous conveyorised systems. Formerly, if the oven attained excessively high product temperatures it was usual to switch off certain heat sources or to substitute lower wattage heat sources, and if too low, to substitute higher wattage heaters. It was also virtually impossible to secure a correct registration of the product temperature, especially if on a conveyorised system. In convection ovens the temperature of the heating air stream is a fair indication of the temperature to which the product will be raised, but in infra-red ovens the air temperature may bear little or no relationship to the product temperature. One method of securing control of conditions is to operate the installation on a time-power density basis which has been indicated by initial laboratory tests or established by production experience.

Present-day practice is to use more sophisticated methods where instant response is necessary and where the importance of close control justifies the expense. Power control in a stepwise manner is the least expensive initially and can vary from manual control to time-switch cycling control. In cycling, the cycles have to be sufficiently short so that temperature swings on the product itself are minor and the overall time cycle of the operation has to allow for several cycles. Where this method of control is inadequate, infinitely variable power control may be necessary. This can be achieved by the use of variable transformers, saturable core reactors, rectifier tubes of the ignitron or thyratron types or of rectifiers of the solid-state type such as a silicon-controlled rectifier. Each type has its advantages and disadvantages and the silicon-controlled rectifier appears to hold the greatest prospect of becoming the accepted method. Such controllers are available up to about 450 kW at the present time. Infinitely variable power control has the great advantage of being capable of being tied with thermocouples, radiometers and other instruments on a feed-back closed loop circuit to allow complete automation of the drying operation, and the savings inherent in automation can frequently far outweigh the initial cost of such controls.

For any such control system to work adequately and accurately it is of course necessary to have a reliable method of ascertaining the product temperature. The use of thermocouples affixed to or embedded in the material undergoing treatment may be suitable for batch operations but is quite unsatisfactory for continuous production lines. Even in batch operations the thermocouple target should be shielded from ambient air conditions within the oven which may be changeable. Excellent temperature measurement can be obtained by means of radiometers which can be focused on to the product and the temperature of which can be read. The most satisfactory are radiometers which are designed to be sensitive in the 5–7 micron region. Being of the photonic type their response is practically instantaneous. It is important that such a radiometer should read the temperature of the product and not energy from the heat source or otherwise reflected. By using radiometers sensitive to the 5–7 micron region, direct energy from the filament of a tungsten lamp, for example, is effectively filtered out, as the glass or quartz envelope has little or no transmission beyond 4·5 microns. The output of the radiometer is fed back into a control panel and can be used to modulate the power so that the product can be maintained at a pre-determined temperature level.

Gas-heated Infra-red Heat Sources

Incandescent filament lamps as a source of infra-red or near infra-red radiant energy were introduced commercially in the late thirties, notably by the Ford Motor Company, and the convenience of their use on assemblies of relatively low first cost led to their increased use for many drying operations, particularly for dry-off applications and the baking of the then new resin-type synthetic metal surface coatings. Poor lamp life and high breakage losses, however, resulted in development of more robust heat sources heated by town gas. Earlier designs took the form of "black-panel" radiant units in which the gas combustion took place through a series of burner jets located behind the panel, the panel itself being raised to a comparatively low temperature in the region of 1,000°F, at which temperature the panel was capable of developing a radiant energy output of around 8,000 Btu/ft^2/h. Such panels possessed long life and were amenable to assembly on suitable framework around conveyorised work in a manner generally similar to electric lamp installations. Meanwhile, electric power energised infra-red heat sources were gradually improved and tended to displace gas-heated panels due to the high power densities attainable and their rapid response to control with little or no residual heat retention. More recently gas-heated sources have been considerably improved and are today capable of extremely high radiant energy output where required. Modern designs of burners are capable of rapid rates of heating-up and cooling, are clean and economical to use and can be rapidly modulated and controlled to most response rates for conveyorised productions. Development of strong porous ceramic plates of light weight have made possible burner surfaces of small area, light weight and considerable strength which can be built into suitable patterns around conveyorised assemblies to give uniform heat distribution to suit the nature and profile of the moving work load. Reflecting surfaces can be incorporated between individual burners to secure maximum utilisation by the work of radiant heat energy emitted from the burner surfaces. Product heating rates can be rapid or slow as desired and product temperature plateaux, once achieved, can be maintained or altered with ease.

The higher the heat source temperature, the higher the radiant heat energy output, but at the same time it is true that, as a general rule, above a certain temperature the higher the heat source temperature, the lower is the efficiency of the flame in heating the source to attain this temperature. Whilst a higher temperature source may be desirable to obtain the greatest radiant heat density, it may not be the most efficient source from the standpoint of greatest heat in the product for the amount of fuel consumed. In the generation of infra-red energy by impingement of a gas flame on a ceramic surface losses will arise from products of combustion which may be lost up the stack or dissipated as convected heat. Such losses will increase with increase in air admitted to the burner beyond that which is required for perfect combustion. The turn-down system employed to achieve heat modulation will usually disturb the gas-air mixture, resulting in excess air supply. The burner surface itself will give rise to unavoidable convection and conduction losses which will increase with increase in temperature. The extent to which the burner flame itself makes useful contact with the ceramic or metal radiating surface will also affect the extent to which it is contributing to raising the temperature of the radiating surface.

Two general types of infra-red burner heat sources are employed at temperatures above "black panel" radiant sources, namely the impingement type and the surface-combustion type. The impingement type is designed to have flame impinging on a ceramic block at relatively high velocity whereby surface temperatures of 2,000°F and above are achieved, with radiant energy generating efficiencies of from 25–40% of the thermal value of the fuel used. This type of burner can attain heat densities in excess of 100,000 Btu/h/ft^2. Surface combustion types of burners for infra-red energy generation operate at lower temperatures and have radiant heat outputs up to 50% of the fuel energy supplied at around 1,560°F colour temperature. The above figures hold for combustion of perfect gas–air mixtures and refer to heat conversion to radiant heat, much of the balance of heat generation being available as hot combustion gases which may be used effectively in heating the work load. Impingement burners would be used where high product temperature, fastest heating rates and shortest cycle times are required. Surface combustion types are most suitable where lower operating costs (higher thermal efficiencies), longer heating cycles and lower heat densities are desired. This latter type is illustrated in Plate XXXVII.

For the relatively low temperatures called for in drying operations as opposed to pure heating purposes the oven will normally be designed with individual burners spaced to some extent from one another, and for the highest utilisation of the radiant energy emitted the intervening spaces will need to be fitted with reflecting surfaces having high infra-red reflecting properties, thereby returning to the material undergoing treatment infra-red energy not initially interrupted and such radiant energy as is reflected from the material itself. A certain amount of the internal wall area of the oven will necessarily have to be sacrificed to permit of removal of products of combustion, convected air and water vapour. The reflecting surfaces will require to be capable of withstanding the conditions within the oven such as temperature and the presence of products of combustion if corrosive. Gold-plated reflectors are not generally satisfactory for this application and anodised aluminium for lower temperatures and stainless steel for higher temperatures are usual. Although any volatiles given off during drying and reaching the burner surfaces will burn off if combustible, interior reflecting surfaces may suffer from accumulations of ash, dust and burnt-on particles

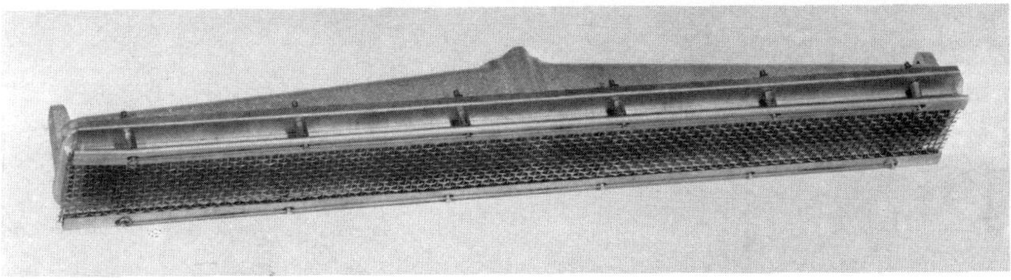

Plate XXXVII. Surface (Schwank) type gas infra-red burner. *Courtesy Fostoria—Fannon Inc.*

and will require periodic cleaning. Suitable safety precautions must also be taken where solvents are evolved during drying, which could lead to explosion or other hazards. The possibility of fuel-gas leakage must also be borne in mind both before and during oven operation, and adequate ventilation must be incorporated into the design so that enough air is moved through the oven to keep concentrations below hazardous levels. In the case of solvent vapours which can lead to explosions the amount of fresh air moved through the oven should be such as to ensure not more than one-quarter the lower explosion limit concentration. Additionally, safety devices must be built into the design to ensure immediate ignition of the fuel when turned on; maintenance of the flame during operation; continuous operation of the ventilating system; and prompt shut-down of fuel supply in the event of failure or stoppage of any part of the system.

Control of temperature of the product during the normal drying cycle as well as in the event of the conveyor system failing—and its subsequent re-starting—will be required, as with electrically energised heat sources. Conveyor interlocks with the fuel supply should be fully automatic. Heat modulation may be manually or automatically operated depending on requirements. As with electric or infra-red ovens, the automatic control can be based on thermocouples or radiometers.

Applications

Infra-red radiant heat is applied extensively today in many heating and baking operations but in the specific field of drying Table 46 gives a good indication of the wide field of its application together with typical drying times experienced.

Freeze Dryers

The use of tray or drawer type batch vacuum drying ovens for heat-sensitive materials or for the drying of materials out of contact with air has been described in Chapter v. Such dryers are usually operated at a vacuum of 27–29·5 in of mercury and at this level of vacuum drying takes

Table 46. Applications of infra-red radiant heat drying

Product		*Drying times*
ADHESIVES	Book binding	5–10 sec
	Laminating tissue to foil	$\frac{1}{2}$–$1\frac{1}{2}$ sec
	Glue on stamps and labels	5–20 sec
	Laminated building panels	1–2 min
	Leather soles	$\frac{1}{2}$–1 min
	Curing brakeshoe bonding	1–5 min
FABRICS	Pre-drying and prevention of dye migration	8–10 sec
	Latex on seat-cover material	$\frac{1}{2}$–1 min
	Drying and curing fibreglass cord	1 min
	Drying fibreglass cloth	2–5 sec
	Drying heavy belting	5–15 min
WOODWORK	Drying sealer coat	$\frac{1}{2}$–2 min
	Drying lacquer	1–3 min
	Drying synthetic paint	5–15 min
METALWORK	Water dry-off	1–3 min
	Lacquer drying	$\frac{1}{2}$–5 min
	Synthetic paint drying	2–10 min

place in a normal manner by the evaporation of water present in the material at the temperature to which it is raised – usually of the order of 90–200°F. Where the product is exceptionally heat-sensitive, and where it is of paramount importance to avoid all thermal deterioration either of chemical composition or of natural properties, as for instance in the preparation of biological, pharmaceutical and therapeutic products, such a method of drying is not adequate. As a result of the demands for antibiotics and blood plasma during the Second World War a method of drying at much lower temperature under very high vacuum was developed and has now attained considerable proportions for the manufacture not only of this type of material but also of edible products, the original qualities of composition, flavour and aroma of which are completely retained in the dried product. This method is known as freeze drying.

Freeze drying consists of freezing the original material and then subliming the frozen moisture content – now as ice – under a vacuum of 100 to 300 microns (0·1–0·3 mm of mercury) at temperatures ranging from −10 to −30°C. The sublimated water vapour is recondensed on refrigerated condensers, adsorbed by a desiccant or removed by a vacuum pump of adequate capacity. The heat of sublimation must, of course be supplied by suitable means either by conduction or by radiation. The heat supply has to be accurately controlled so that on the one hand sublimation does not cease and on the other the material does not melt during the drying operation. Sublimation of the ice commences at the surface of the material and the dry surface layer so formed remains static in view of the fact that the whole of the material is frozen. As drying progresses the ice surface recedes gradually within the body of the material, leaving a porous structure in the dried product through which water is free to move to the external surface. The rate of drying remains substantially constant throughout this period. As the ice layer gradually recedes through the material and only sublimed water vapour moves to the surface, to the exclusion of liquid moisture, concentration of dissolved salts or components at the surface which occurs in normal vacuum tray drying is precluded and the product retains its original composition of dry matter uniformly dispersed. All the original constituents have remained frozen in their original positions within the material and in consequence the dry product will usually retain its original shape and size and will have considerable porosity, resulting in subsequent ready solubility or dispersion in water. This is of major importance in the manufacture of penicillin, dried blood plasma, fruit juices, coffee and other products.

The drying operation is usually carried out in two stages. During the first stage the ice in the frozen material is sublimed by the action of the vacuum and a minimal heat supply, the temperature of the material being maintained well below freezing point. During this stage it is usual to remove 98–99 % of the contained moisture. The supply of heat controls the movement of the ice layer inwards and it has to be controlled in such a manner as to yield the highest possible water vapour pressure at the ice surface without melting the material; the rate of heat supply is critical. In the second stage of drying the heating is less critical and residual moisture is comparatively readily released due to the porous nature of the product, although the removal of this last 2 % of moisture may take a considerable time. The material is allowed to attain room temperature or even higher, to secure moisture contents below 0·5 %, and this can usually be safely permitted as the material is too dry to melt when it is warmed. Florsdorf[24] found that in drying sera containing 10 % solids at -18°C the ice surface receded at the rate of 1 mm/h and that drying proceeded at a constant rate until all the ice had been removed. It took 80 % of the total drying time to remove 95 % of the total moisture; 90 % of the total drying time to attain 1 % residual moisture; and the remaining 10 % of the drying time to reduce the moisture content to 0·5 %. The drying rate is affected by the thickness of material – the thinner the layer the higher the drying rate – and the nature of the product. Protein solutions may, for instance, take eight to ten times as long as pure water. For this reason it may sometimes be desirable to pre-concentrate by normal vacuum tray drying before freeze drying unless a very highly porous product is required. Optimum throughputs are usually obtained with material depths or thicknesses of $\frac{1}{2}$–$\frac{3}{4}$ in. Freeze-drying plants are usually operated at temperatures below the eutectic point of the product and the lower this is the longer is the drying time. Miner[25] suggests that the higher the sugar content of the product the lower is the eutectic point, and hence longer drying times are required for products of increasing sweetness. Freeze-dryer chambers are operated at pressures below that corresponding to the eutectic point of the product in order to improve the quality of the product, provide a margin of safety against melting and to allow for more rapid heat input to shorten drying cycles.

The essential components of a vacuum freeze-drying plant are: (a) the chamber or chambers in which the material is loaded on trays or other containers and provided with a heat supply either in the form of radiant panels or warm heating medium coils or platens; (b) a moisture vapour condensing or adsorption system; and (c) evacuating equipment in the form of vacuum pumps or ejectors. Each drying chamber may be provided with its heating, refrigerated condensation and evacuating equipment or the chambers may be disposed in groups hooked-up to a reduced number of condensation and evacuating systems. As the process is usually batchwise and has to cope with a widely ranging cyclical operation of the refrigeration and vacuum equipment, staggered or batch-cycling operation of the chambers in groups has advantages and economies. A typical plant will have a central roughing pump or ejector for initial large capacity development of vacuum to approximately 1 or 2 mm mercury and a holding vacuum pump for each cabinet or chamber to operate through about three-quarters of the cycle. Each chamber would also have its own holding vapour condenser which can be defrosted during turn-round periods and the initial evacuating period. A large central vapour condenser is operated to take the peak water vapour load during the first quarter of the drying cycle and to alternate on a staggered load basis between chambers. A central refrigeration plant would be employed to deal with a stable load and a more constant evaporator temperature. Normal cycles of operation would consist of:

1. Loading of the chamber with raw material in suitable containers already frozen (to save cycle time) and with the already defrosted vapour condenser valved-off from the chamber.
2. Putting the refrigeration system into operation to chill and dry coils at zero load.
3. Closing the chamber, starting up the vacuum system for maximum initial load and opening the vapour condenser valve. The system is run at the low temperature prevailing to initiate ice evaporation.
4. Increase of heat input to chamber coils, plates or radiant panels,

inducing high evaporation rates and hence increased refrigerated condenser loadings; chamber and material temperatures undergo fairly rapid increase during this stage. Biological products are usually supported in trays placed on heated shelves and the present-day practice is towards arranging an air space between trays and shelves in order to heat by convection and radiation rather than by direct conduction. Foodstuffs, on the other hand, may have radiation panels which may be heated to 200–300°F at the initial stages and later reduced to 125–150°F as the vaporisation rate falls off. Heating by conduction for this application may lead to burning of the food product if the temperature is not maintained at a low level involving excessively long drying periods.

5. At a pre-determined point for a given product decreasing the heat input to prevent product damage.
6. Where more than one condenser is used, a switch-over to a chilled defrosted condenser is made whilst the one already in use is defrosted. Sometimes several condenser coils are used in order to stabilise operational loadings which are inherently other than smooth and to stabilise temperature conditions. The last hour of the complete cycle will have very light loads both as regards evacuation capacity and heat input.
7. Cooling of the product. It is normal to remove the product whilst still warm so as to avoid moisture pick-up from the atmosphere. Pharmaceutical and biological products are usually dried in ampoules or bottles which can be sealed immediately on removal from the dryer, thus protecting the natural hygroscopic nature of the highly porous product. Foodstuffs and fruit juices require to be packed and sealed as quickly as possible after cooling in sealed gas-tight chambers.
8. Breaking the vacuum by introduction of an inert gas where necessary. In the case of mushrooms, air is used.
9. Unloading to a low humidity store room when immediate sealing of containers or packaging is not called for.

The freezing of the material for stage (1) could be carried out in the chamber itself (auto-freezing) but this would normally be considered too extravagant a use of high capital cost equipment. It is more usual to submit the product to freezing to about −30°C in a separate plant using a cold air blast freeze. The auto-freezing method may also result in loss of moisture and consequent shrinkage of the product, which may be undesirable. Whereas mechanical vacuum pumps would usually be used for therapeutic, biological and similar products, where adequate water supplies at around 60°F are always available steam ejectors would be used for foodstuffs and fruit juices and coffee. Steam jets are, however, less adaptable than vacuum pumps as they are usually less stable at pressures other than those for which they were designed. Mechanical pumps on the other hand cannot handle the very large vapour volumes evolved at the existing pressures and depend upon the effectiveness of the refrigerated condensers for their satisfactory operation. The degree of vacuum can in fact be controlled by the setting of the refrigerant supply valve. It must be remembered that the specific volume of water vapour at, for instance, $\frac{1}{2}$ mm mercury pressure is 24,000 ft^3/lb and this would be equivalent to about 16,000 ft^3/min for a plant having a capacity of 1,000 lb moisture removal per 24 h day.[25] For every pound of moisture removed the heat requirements are about 12,000 Btu and the refrigeration load will be of the same order.

Refrigerated condensers operated at low enough temperatures remove water vapour completely so that after initial evacuation the pumps will only have to remove small amounts of permanent gases. The heat-transfer coefficient from water vapour to an ice surface is low and the heat conductivity of ice itself is low. Hence condensing surfaces have to be very large and defrosting of surfaces has to be resorted to. As freeze drying is generally operated at temperatures well below the triple point of ice, water and water vapour condensation is in the form of hoar frost which has a particularly low conductivity. The drying cycle will be dependent on the difference between the vapour pressure of the receding ice surface in the material and the ice surface on the condensers. Too thick a layer of ice on the condenser surfaces may mean that the surface may be many degrees higher than the refrigerant temperature within the condenser and may result in loss of operating efficiency and plant capacity. Hence it is customary to use mechanical scrapers on the condensers or to use rotary condensers and scraper bars to remove ice during actual operation. On some plants the released hoar frost drops into a container fitted with a mechanical ram system which ejects the ice in block form without loss of vacuum.

Vapour pressures over pure ice are given in Table 47.

Only ice vapour will exist at temperatures below 32°F or at pressures below 4·6 mm mercury absolute. For a given product temperature in freeze drying there is a minimum condenser temperature below which little is gained by further refrigerant temperature reduction. Few products are dried at temperatures below −30°C.

Commercial freeze-drying plants are today supplied in standard sizes having 300, 500, 700 and 1,000 ft^2 of shelf area. Occasionally plants are installed having 1,500 ft^2 of shelf area. Two and a half pounds of raw material (foodstuffs) per square foot is quite usual. Cycle times can vary from 8 h for diced chicken, 9 h for Brussels sprouts, 10 h for mushrooms and 12 h for shrimps[25], depending among other factors on moisture

Table 47. Vapour pressure over ice

Temperature		*Pressure*			
°*F*	°*C*	*lb/in*^{2} *abs*	*in Hg*	*mm Hg*	*Microns*
32	0	0·089	0·18	4·6	4,600
30	−1	0·081	0·16	4·2	4,200
20	−7	0·051	0·10	2·6	2,600
10	−12	0·031	0·06	1·6	1,600
0	−18	0·019	0·04	1·0	960
−10	−23	0·011	0·02	0·6	560
−20	−29	0·006	0·01	0·3	320
−30	−34	0·004	0·007	0·2	180
−40	−40	0·002	0·004	0·1	96

content. Chicken meat for example has 56% moisture content against 90% for mushrooms. Total processing costs vary in the same way, being of the order of 4·3 cents per pound for chicken against 5·8 cents for mushrooms for a given size of plant. The cost decreases with increase in plant capacity. A 4 ton/24 h day dried product capacity is quite normal but plants upwards of four or five times this size have also been installed.

For larger productions of freeze-dried foodstuffs continuously operated plants have been installed consisting of long cylindrical tunnels equipped with entry and exit airlocks. Vapour condensers and vacuum off-takes are grouped along the tunnel length to suit the vapour load at different sections of the tunnel corresponding to the requirement at any stage of the drying cycle. Heat supply to the platens or plates is similarly automatically controlled.

Quite apart from the essential nature of this type of plant for biological and therapeutic products, freeze drying has great attractions in the prepared and pre-packed food industry. Freeze-dried foods have long shelf life without refrigeration—two years for a 2% residual moisture content product being usual. Such a product may weigh from 10–50% of the same product in its raw, frozen or canned form resulting in economies in handling and shipping. The product is equally as fresh after a period of time as when in season and may frequently be of superior flavour than when processed in any other way. Foods having a high fat content and particularly high sugar content are not very suitable for freeze drying and in general require to be of initially high cost to justify this method of processing. Freeze-dried coffee, although more costly than the spray-dried product, is claimed to be of superior flavour and aroma. Vast quantities of fruit juices are today dried by this method. Freeze drying is now standard practice for penicillin, streptomycin, protein hydrolysates, hormones, blood plasma, sera, many vitamin preparations, certain yeasts and moulds.

Dielectric Dryers

Forced convection, conduction and to a large extent infra-red radiant heat dryers suffer the disadvantage that heat has to be transmitted through the moist material from the surface inwards, mostly by conduction. Moisture migration to the surface depends upon the temperature gradient from the surface to the centre in the case of heating from all sides, or from the exposed surface through the total thickness of material in the case of heating one exposed surface only. In dielectric or high-frequency drying the whole mass of material is raised in temperature more or less uniformly as a result of high-frequency changes of stress on the molecules of the material subjected to treatment. Limitations of surface film coefficients and internal heat conductivity do not arise. Basically the elements of a dielectric dryer are composed of two parallel plates between which the material rests or is conveyed, each plate being connected to the terminals of a high-frequency electronic generator. The system represents a capacitor, the moist material being the dielectric of the capacitor. The two plates are alternately connected to positively and negatively charged impulses which set up reversing stresses in the molecules at each half-cycle of the electric supply. The reversal of the polarity of the plates and hence the change in stress in the molecules corresponds to the frequency of the electronic generator which will lie in the range of 2×10^6 to 100×10^6 c/s.

This violent and high-frequency change in stress on the material molecules generates heat within the material and, as the electrical field creating this phenomenon is uniform throughout the thickness of the material, development of heat therein is substantially uniform. Temperature development at the centre of the material layer will therefore be, for all practical purposes, the same as at the surface, a condition conducive to rapid and uniform drying of the whole, assuming moisture vapour is free to migrate to and from the material surface. We have the situation therefore that migration of moisture to the surface is not dependent on a temperature gradient from the surface inwards nor on the conduction of heat through the thickness of the material; moreover internal heat may be developed extremely rapidly which, assisted by purging of moisture vapour from the surface by convected air, will lead to rapid uniform drying. Capital cost of equipment is, however, high as are power requirements.

The theoretical power requirements for dielectric heat drying are given approximately by the expression:

$$P \text{ (in kW)} = 2{\cdot}93Mc_p(t_2-t)\times 10^{-4}+\frac{W}{3}$$

where M = feed rate lb/h dry material

c_p = specific heat at constant pressure of feed, Btu/lb

Plate XXXVIII. Continuous dielectric core drying oven.
Courtesy FECO, a division of Bangor Punta Operations Inc.

W = evaporation, lb/h

$t_2 - t$ = rise in temperature of material, °F

The theoretical power requirement should be increased by 20–25% to allow for conduction, convection and radiation losses and in addition allowance must be made for the overall efficiency of the dielectric heating installation which will usually be of the order of 50–60%. Nevertheless, due to the rapidity with which the wet material temperature can be raised, drying times are materially shortened where this method is applicable and and can show substantial savings over more conventional methods of drying. Published performance figures in dielectric drying are extremely meagre but the saving in drying time in the case of grinding wheels is well illustrated in Table 48. Dielectric drying can be seriously considered as a rule where the percentage of moisture to be removed is low, where the removal of this moisture by conventional methods is a slow operation and where the final product is of relatively high intrinsic value. In general, moisture removal is found to be at the rate of $1\frac{1}{2}$–2 lb/kWh consumed.

Important reductions in drying times of this order result in considerable saving in factory floor space occupied by more conventional drying equipment or may result in substantial increase in capacity of continuous

Table 48. Comparison of drying times for grinding wheels by convection and dielectric methods

Grinding wheel size	*Convection drying*	*Dielectric drying*
700 lb × 12 in thick	5 days	70 min
450 lb × 8 in thick	3 days	45 min
330 lb × 6 in thick	$1\frac{1}{2}$ days	30 min
100 lb × 3 in thick	8 h	12 min

Courtesy Radyne Ltd.

baking ovens processing rusks, dog biscuits, cookies and similar products where the drying is carried out in a dielectric unit instead of within the oven. Although dielectric heating systems are more widely employed on actual heating operations, they find suitable drying applications, in addition to the above, to breakfast cereals, starch-reduced rolls, furniture timber blanks, veneers and plyboard, plasterboard, water-based foamed plastic slabs and certain textile products.

CHAPTER XIV

The Drying of Gases

Many industrial processes call for a supply of air of a lower humidity than that which is conveniently and economically attained by compression or chilling. For normal ranges of air conditioning adequate dehumidification can be accomplished by chilling, but many special processes call for much lower relative humidities or closer control of humidity than can be satisfactorily obtained by this method. Low or negligible humidities may be required in the manufacture or thermal drying of certain foodstuffs, gelatines and confectionery or for special storage facilities, precision control of electrical equipment and automatic telephone exchanges. Both organic and inorganic gases may require drying in order to meet market or process requirements. Two methods are used to attain low-level relative humidities, namely absorption and adsorption.

Drying by absorption may be accomplished by the use of either liquid or solid desiccants. Liquid desiccants are generally organic liquids, concentrated acids, alkalis or salts in solution; they remain liquid throughout the process. They can be used in spray chambers or in packed columns and accomplish dehumidification of the gas by chemical combination to form hydrates or otherwise by absorption and solution. Being in liquid form they can be conveniently applied as a continuously operating drying process. Typical liquid desiccants are sulphuric acid, lithium chloride, calcium chloride, potassium hydroxide and caustic soda solutions; glycerol, diethylene glycol and triethylene glycol. Solid desiccants, which remain solid throughout the drying process, are generally used in equipment designed to operate intermittently with provision for periodic regeneration. The two most generally used solid desiccants which remove moisture by chemical combination are anhydrous calcium sulphate and lime.

Liquid desiccants, which are normally anhydrous, could be used to attain minimal gas humidities but are rarely used in this state due to the cost of re-dehydration and loss of material. Regenerative temperatures may also be so high as to result in complications. It is usual to use them in concentrated solution form in the same manner as with deliquescent salts and hydrates. In general, such desiccants must possess in solution a low water-vapour pressure, and should be non-corrosive, maintain the desiccating action over a wide range of concentration, remain stable during the drying process and during regeneration, be non-toxic, and of low cost.

Both acids and alkalies are widely used in industry in spite of their undesirable handling characteristics, but the most used today are the glycols and lithium chloride.

Liquid desiccants have a considerable capacity for moisture absorption and for this reason have advantages when very considerable volumes of moist air or gases have to be dried. The relative capacities to those of solid adsorption desiccants are shown in Fig. 102, where the equilibrium moisture content of the desiccant is plotted against the percentage relative humidity of the air. However, under practical conditions liquid desiccants are not very suitable for the attainment of very low level relative humidities and are not generally used to obtain humidities below 20–30%. Solid adsorption desiccants on the other hand are easily capable of obtaining

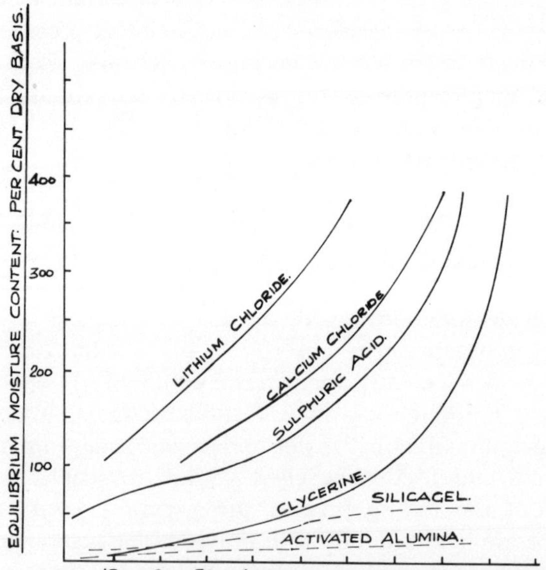

Fig. 102. Relative capacities of solid and liquid desiccants.

commercial dryness, usually accepted as corresponding to a dew-point of −40°F or even lower.

Fig. 103 gives the saturation water-vapour content of air over the range normally covered by air-drying installations.

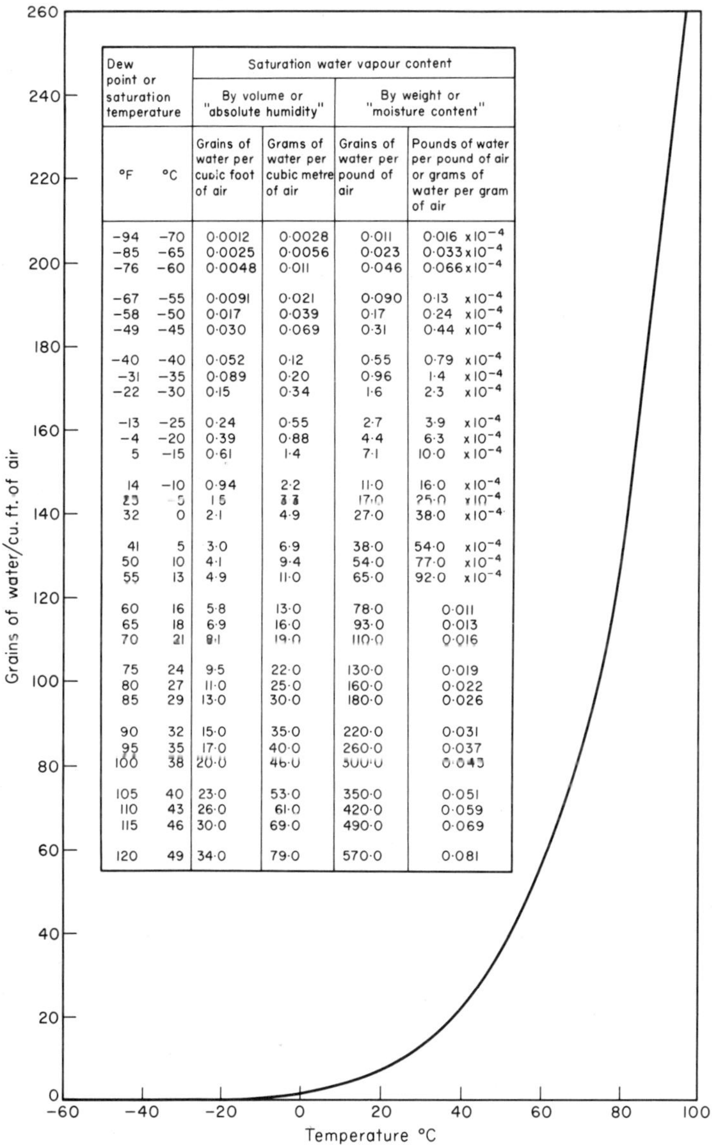

Dew point or saturation temperature		Saturation water vapour content			
		By volume or "absolute humidity"		By weight or "moisture content"	
°F	°C	Grains of water per cubic foot of air	Grams of water per cubic metre of air	Grains of water per pound of air	Pounds of water per pound of air or grams of water per gram of air
−94	−70	0·0012	0·0028	0·011	0·016 ×10⁻⁴
−85	−65	0·0025	0·0056	0·023	0·033 ×10⁻⁴
−76	−60	0·0048	0·011	0·046	0·066 ×10⁻⁴
−67	−55	0·0091	0·021	0·090	0·13 ×10⁻⁴
−58	−50	0·017	0·039	0·17	0·24 ×10⁻⁴
−49	−45	0·030	0·069	0·31	0·44 ×10⁻⁴
−40	−40	0·052	0·12	0·55	0·79 ×10⁻⁴
−31	−35	0·089	0·20	0·96	1·4 ×10⁻⁴
−22	−30	0·15	0·34	1·6	2·3 ×10⁻⁴
−13	−25	0·24	0·55	2·7	3·9 ×10⁻⁴
−4	−20	0·39	0·88	4·4	6·3 ×10⁻⁴
5	−15	0·61	1·4	7·1	10·0 ×10⁻⁴
14	−10	0·94	2·2	11·0	16·0 ×10⁻⁴
23	−5	1·5	3·3	17·0	25·0 ×10⁻⁴
32	0	2·1	4·9	27·0	38·0 ×10⁻⁴
41	5	3·0	6·9	38·0	54·0 ×10⁻⁴
50	10	4·1	9·4	54·0	77·0 ×10⁻⁴
55	13	4·9	11·0	65·0	92·0 ×10⁻⁴
60	16	5·8	13·0	78·0	0·011
65	18	6·9	16·0	93·0	0·013
70	21	8·1	19·0	110·0	0·016
75	24	9·5	22·0	130·0	0·019
80	27	11·0	25·0	160·0	0·022
85	29	13·0	30·0	180·0	0·026
90	32	15·0	35·0	220·0	0·031
95	35	17·0	40·0	260·0	0·037
100	38	20·0	46·0	300·0	0·043
105	40	23·0	53·0	350·0	0·051
110	43	26·0	61·0	420·0	0·059
115	46	30·0	69·0	490·0	0·069
120	49	34·0	79·0	570·0	0·081

Fig. 103. Saturation water-vapour content of air at various temperatures.

Courtesy A.E.I. – Birlec Ltd.

Adsorption desiccants such as silica gel and activated alumina remove moisture from gases by virtue of the fact that moisture vapour adsorbed into the extended internal areas of the material by capillarity and condensation as a monomolecular layer on these surfaces is at less than the partial pressure of the water vapour in the air or gas stream. Desiccation of the gas continues until an equilibrium condition between the adsorbed moisture and that in the external gas is established. The capillary space in silica gel, for instance, occupies about 50% of the specific volume and a high degree of air dehydration is possible until the silica gel has adsorbed up to 20% of its own weight of water. It is physically and chemically stable and can be regenerated by the application of heat at temperatures below 500°F an unlimited number of times. The residual moisture after re-activation is about 5% by weight and about 2,000 Btu/lb removed moisture is required for regeneration. Its bulk density is 40–45 lb/ft^3 depending upon mesh size which usually ranges from 6–16 mesh. It adsorbs water vapour preferentially in the presence of other vapours such as solvents and hydrocarbons. The pressure drop through a bed of silica gel at different air-flow velocities is given in Fig. 104.

Activated alumina, prepared from aluminium trihydrate, is an inert, stable, non-corrosive granular material possessing considerable porosity. It is capable of removing substantially all the moisture content of a gas until it has adsorbed about 14% of its own weight. It will continue to remove moisture at a lower level of efficiency until its moisture content reaches

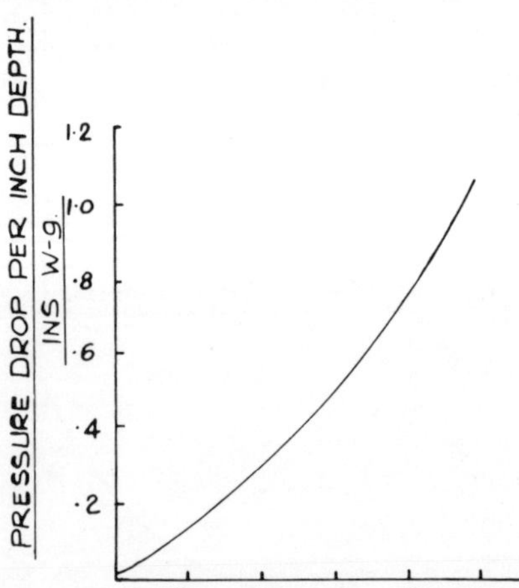

Fig. 104. Pressure drop through silic gel bed.

(after P. Brennan. The Heating & Ventilating Engineer. Feb. 1963)

40% of its weight. Reactivation, as in silica gel, is accomplished by heating to below about 400–450°F. In both cases, this is usually done by an externally heated source of hot air although additional heating elements or coils may be employed in the bed of desiccant. Activated alumina impregnated with calcium chloride, which has about twice the adsorptive capacity of unadulterated alumina, should not be regenerated at temperatures above 350°F. Normal mesh size used for drying plant beds has a bulk density of about 50 lb/ft^3. Pressure drop through the bed depends on mesh size and is given in Fig. 105 for the most usually employed range.

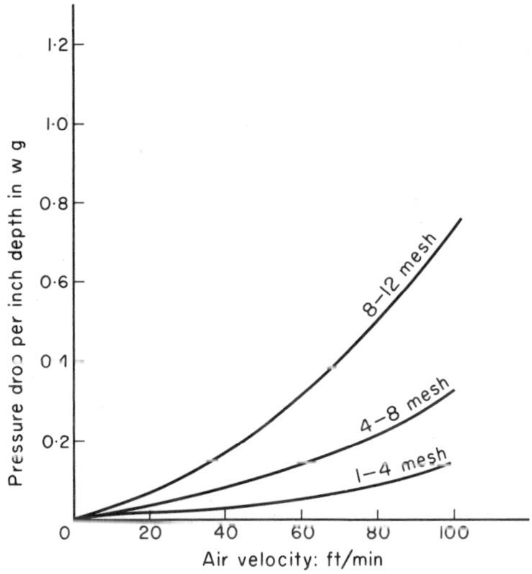

Fig. 105. Pressure drop through activated alumina beds.

The heat of moisture adsorption generated in the bed of desiccant is considerable. Maximum bed temperatures occur at the level in the bed where moisture adsorption efficiency is at a maximum and this level will move progressively through the height of the bed from moist air inlet to dry air discharge. The exit air temperature will remain relatively low during the operation of the drying bed for some time but will then commence to rise, reaching a maximum when the highest efficiency adsorption level reaches the termination of the bed–the top of the bed in bottom air inlet operation. It is not unusual to maintain the desiccant bed at a reasonably low level by the use of immersed cooling coils, but even so the exit air temperature will pass through a maximum, at which stage it will be known that the bed as a whole will have accomplished its drying to the maximum degree of efficiency. Experience will dictate the point at which the bed will be taken off stream and subjected to the regeneration stage. Alternatively, the switch-over to the regenerative cycle will be actuated by an automatic hygrometric controller in the exit gas. After reactivation by means of a stream of steam or heated air, the bed requires to be cooled by cold dry air or by means of cooling coils embedded in the desiccant. Fig. 106 illustrates a typical flow diagram of an adsorption dryer. No. 1 cylinder is fed from a blower or fan with the air or gas to be dried, usually in an upward direction, whilst No. 2 cylinder undergoes reactivation by steam or heated air, usually in a downward direction. The equipment may be designed so that the drying cylinder takes care of the throughput of gas required for one working shift, No. 2 cylinder usually taking the same time to regenerate and to be cooled.

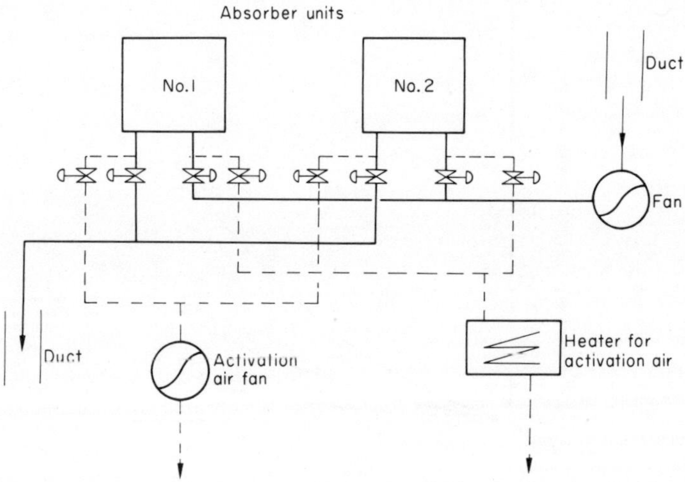

Fig. 106. Typical flow-system for dual bed automatic absorber. *Courtesy A.E.I. – Birlec Ltd.*

Switch-over from drying to regeneration may be manual for small units based on a time cycle or temperature change in the exit gases, or in larger plants by thermostatic control actuated by the temperature change or by hygrometric controls based on humidity of the exit gas through motorised valves. Alternatively, the switch-over valves are actuated by a programme controller system based on predetermined time cycles.

For reliable practical design purposes, the procedure for the determination of the adsorptive capacity of desiccants such as silica gel and activated alumina is somewhat complex. The adsorptive capacity under isothermal conditions is first calculated. It is then assumed that in practice adsorption will be under adiabatic conditions and the effect of the resultant rise in temperature of the desiccant bed is to lower the humidity of the bed. Then, by calculating the temperature rise in the gas stream and deriving the relative humidity over the bed, it is possible to use the isothermal data previously determined to estimate the adsorber capacity. However, the approximate quantity of adsorbent required for a particular duty can be fairly easily calculated. Thus, assuming 100 ft^3/min of air having a humidity of 5·5 grains water/ft^3 of dry air has to be dried down to 0·5 grains/ft^3 and that the dryer is to be capable of operating one of its two cylinders over an eight-hour shift between regenerations, the quantity of water to be removed in the eight-hour period will be

$$\frac{8 \times 100 \times (5{\cdot}5 - 0{\cdot}5) \times 60}{7{,}000} = 33{\cdot}6 \text{ lb}$$

Assuming that silica gel is to be used, it would be safe to expect that it will adsorb 20% of its weight of water and that it can be regenerated to a 5% residual moisture content. The moisture removal capacity will therefore be 15% of its weight. The weight of silica gel required will therefore be

$$\frac{33{\cdot}6}{0{\cdot}15} = 224 \text{ lb}$$

Assuming a superficial face velocity of the entering air at 60 ft/min, a flow of 100 ft^3/min will therefore require a minimum cross-sectional area of the desiccant bed of 1·67 ft^2, or for practical purposes 2 ft^2. Taking the bulk density of silica gel as 45 lb/ft^3, the desiccant bed would require to be 5 ft^3. Each drying cylinder would therefore require to be 20 in in diameter and allow 2 ft 6 in for the bed height, assuming a single-layer bed involving a pressure drop through the bed, from Fig. 104, of 0·5 in w.g./in of bed depth, or say a total of 15 in w.g.

Such a calculation is a reasonable approximation for pressures of a few pounds per square inch, but when dealing with gases at high pressure corrections have to be made to allow for the fact that the perfect gas law ($PV = RT$) does not apply. A correction factor known as the Compressibility Factor has to be applied to take care of the fact that the atoms and molecules of the gas themselves occupy a certain volume. Compressibility factors for air, taken from International Critical Tables, are given in the form of curves in Fig. 107 with air at different temperatures over a range of pressures.

Thus, for air flowing at 250 atm pressure and 15·7°C, a factor of 1·12 has to be applied since the actual volume of air flowing is about 12% higher than that calculated on the relationship $PV = RT$. Furthermore, due to the increased density of the gas at high pressure, the gas behaves rather like a liquid and the solubility of moisture in the compressed gas as opposed to saturation moisture content assumes importance. Dr. T. J. Webster(28) of the British Oxygen Company from research in this field established the relationship between pressure and superimposed pressure,

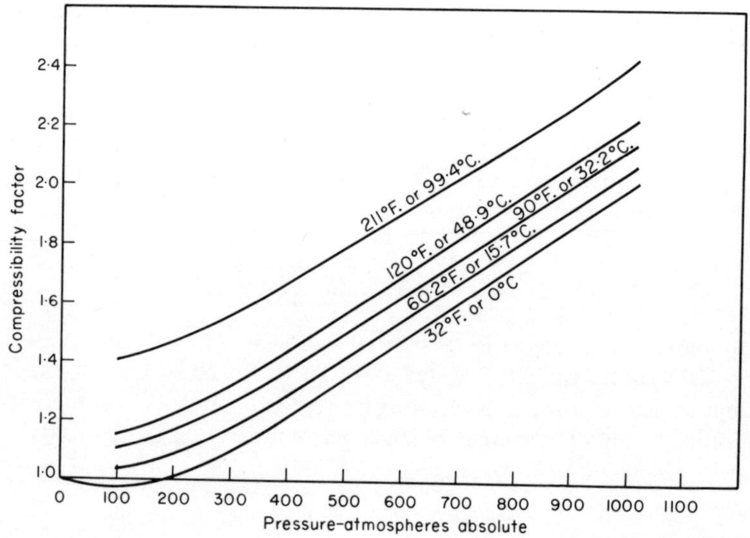

Fig. 107. Compressibility factors for air. *(After International Critical Tables.)*

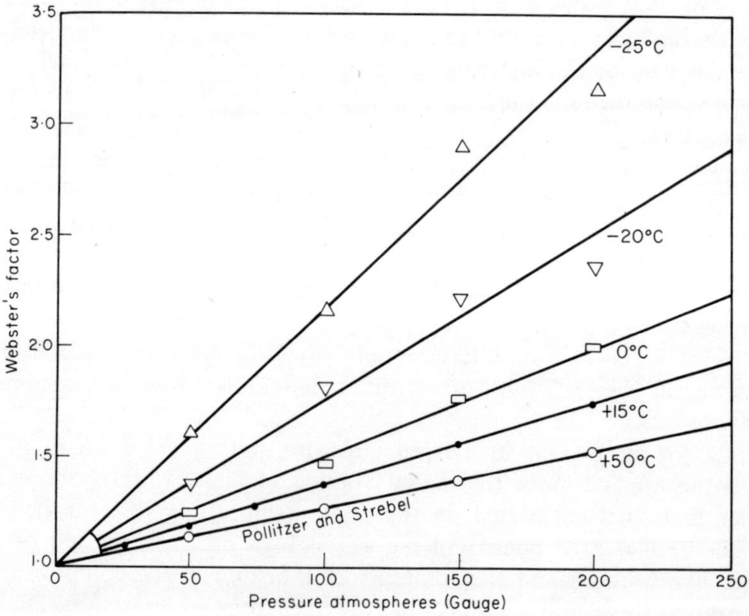

Fig. 108. Relationship between pressure and super-imposed pressure for air. *(After Webster, T.J., J.S.C.I., Nov. 1950.)*

for air resulting in an empirical correction known as Webster's Factor as shown in Fig. 108 at different temperatures.

Thus, assuming 750 s ft^3/min of air saturated at 16°C has to be dried in an adsorber over an 8-hour cycle at a pressure of 350 atm. the flowing cubic feet of air will be 750/350 = 2·15 ft^3/min at pressure or 1,030 flowing cubic feet in 8 hours. Air saturated at 16°C contains 5·8 grains moisture per ft^3 of air so that the moisture load on the dryer will be $1{,}030 \times 5{\cdot}8/7{,}000 = 0{\cdot}854$ lb in 8 hours. The compressibility factor is 1·12 and Webster's factor from Fig. 108 is 1·95, giving a corrected moisture load in 8 h of $0{\cdot}854 \times 1{\cdot}12 \times 1{\cdot}95$ or 1·87 lb, which is rather more than double the load anticipated were the perfect gas law to apply, and the adsorbent capacity of the dryer would have to be increased accordingly.

The adsorption desiccants most frequently met with in industrial practice, namely silica gel and activated alumina, have been treated in some detail, but as mentioned earlier a wide range of desiccant materials are used, and Table 49[(28)] lists some of these and gives their residual moisture contents in mg/litre and the equivalent dew-point of air in contact with them.

Table 49. Comparative effectiveness of desiccants

Material	*Residual moisture content mg/litre*	*Equivalent dew-point °F*
Copper sulphate (anhydrous)	2·8	+ 19·4
Calcium chloride (granular)	1·5	+ 10·4
Calcium chloride (tech. anhydrous)	1·25	+ 1·4
Zinc chloride (sticks)	0·98	− 2·2
Caustic soda (sticks)	0·80	− 5·8
Calcium chloride (anhydrous)	0·36	− 20·2
Potassium hydroxide (sticks)	0·014	− 68·8
Calcium sulphate (anhydrous)	0·005	− 83·2
Calcium oxide	0·003	− 88·6
Sulphuric acid (95% at 35°C)	0·002	− 94·0
Barium oxide	0·00065	− 106·6
Molecular sieves	0·00065	− 107

Those salts and hydrates which are deliquescent are in practice, as has been explained, used in concentrated solution. Calcium chloride would usually be used as a 40–50% aqueous solution, whilst sodium and potassium hydroxides would be used as saturated solutions at the operating temperature. Anhydrous calcium sulphate, on the other hand, in its granular form has sufficient mechanical strength to permit its use as a desiccating bed, forms the hemihydrate $CaSO_4\frac{1}{2}H_2O$ on adsorption of moisture, containing only 6·6% of its weight of water, which is a limitation on its use. However, it has a highly porous structure which promotes adsorption to a total capacity of 12–14% of its weight in saturated air. Dew-points down to

$-80°F$ are possible and its drying efficiency is maintained to near exhaustion at temperatures below 150°F. It is an excellent desiccant for organic liquids with reasonably low boiling points (up to 210°F), is regenerated at temperatures of 400 to 425°F and will withstand about 200 regenerations.

Magnesium perchlorate, $Mg(ClO_4)_2$, is a highly efficient desiccant having residual moisture of 0·031 mg/litre with an equivalent dew-point of $-58°F$, will withstand a temperature rise to 275°F without appreciable loss in efficiency, and when saturated to the form of hexahydrate, $Mg(ClO_4)_2 6H_2O$, has a content of 48·6% of its dry weight as moisture. Loss of drying efficiency occurs as each subsequent hydrate, the di-, tri- and hexahydrate, is reached; even at the trihydrate stage it equals the desiccating power of sodium or potassium hydroxide in the solid form. It requires a temperature of 400–500°F under high vacuum for complete regeneration which is a disadvantage.

Barium oxide is highly active as a desiccant up to 1,000°F but regeneration, in turn, calls for higher temperatures, which make its use generally impracticable. Calcium oxide is, on the other hand, frequently used, particularly because of its low initial and replacement cost.

Sulphuric acid has for many years been used as a drying agent in the chemical industry and, where it can be conveniently handled, is highly efficient. Due to the practical dangers in its use, however, it has been largely displaced by other desiccants.

Most of the afore-mentioned desiccants show a preferential activity towards moisture and may therefore be used with organic gases. Activated carbon, on the other hand, although an extremely active desiccant, has a selective action, showing a preference for higher molecular weight gases and vapours. It is therefore normally used for solvent and vapour recovery, any retained moisture being separated from a non-miscible hydrocarbon on regeneration by steam or recondensation.

APPENDIX I

Conversion Factors

Area	1 m^2	= 1,550 in^2 = 10·76 ft^2
Volume	1 ft^3	= 28·32 litres = 0·02832 m^3 = 6·24 Imp. gal
	1 m^3	= 35·31 ft^3 = 221 Imp. gal
	1 Imp. gal	= 4·536 l = 1·201 U.S. gal
	1 litre	= 0·22 Imp. gal = 0·264 U.S. gal
Velocity	1 cm/s	= 0·03281 ft/s = 1·9685 ft/min
	1 ft/min	= 0·01667 ft/s = 0·508 cm/s
	1 ft/s	= 30·48 cm/s
Mass	1 kg	= 2·205 lb = 15,432 gr
	1 lb	= 0·4536 kg = 7,000 gr
Pressure	1 torr (mm Hg abs.)	= 0·00136 kg/cm^2 = 0·0193 lb/in^2 = 0·00131 atm
	1 kg cm^2	= 735·56 torr = 14·223 lb/in^2 = 0·967 atm
	1 lb/in^2	= 51·714 torr = 0·0703 kg/cm^2 = 0·068 atm
		= 2·036 in Hg = 27·7 in H_2O = 735·5 mm Hg
	1 atm	= 760·0 torr = 1·0332 kg/cm^2 = 14·696 lb/in^2
		= 29·92 in Hg = 33·92 ft H_2O
	1 mm Hg	= 0·0394 in Hg = 0·535 in H_2O
	1 in Hg	= 25·4 mm Hg = 13·595 in H_2O
	1 in H_2O	= 1·868 mm Hg = 0·0735 in Hg
	1 ft H_2O	= 62·381 lb/ft^2 = 0·4332 lb/in^2
		= 0·03045 kg/cm^2
Power	1 W	= 0·00134 hp = 0·102 kg m/s
	1 hp	= 745·7 W
	1 kg m/s	= 9·8 W = 0·013 hp
Heat	1 Btu	= 253 g cal = 1055 J = 778·1 ft lb
		= 0·555 Ch (Cheval vapeur or metric horsepower)
	1 kWh	= 3,415 Btu = 860,445 g cal
	1 kcal	= 3·968 Btu = 2·205 Ch
Heat flow	1 Btu/ft^2/h	= 0·2712 g cal/cm^2/h
		= 2·712 kcal/m^2/h
	1 kcal/ft^2	= 0·3687 Btu/ft^2/h
Heat transfer	1 Btu/ft^2/h/deg F	= 4·882 kcal/m^2/h/deg C
	1 kcal/m^2/h/deg C	= 0·2048 Btu/ft^2/h/deg F

APPENDIX II

Evaporation per ton of Wet Feed or Dried Product

% *moisture content of wet feed*	% *moisture content in dried product*							
	1	5	10	15	20	25	30	35
5	94							
10	224	124						
15	369	264	132					
20	532	420	280	140				
25	717	597	448	299	149			
30	928	800	640	480	320	160		
35	1,172	1,034	862	689	517	345	172	
40	1,456	1,307	1,120	933	747	560	373	187
45	1,792	1,629	1,425	1,222	1,018	815	611	407
50	2,195	2,016	1,792	1,568	1,344	1,120	896	672
55	2,688	2,489	2,240	1,991	1,742	1,493	1,244	995
60	3,304	3,080	2,800	2,520	2,240	1,960	1,680	1,400
65	4,096	3,840	3,520	3,200	2,880	2,560	2,240	1,920
70	5,152	4,853	4,480	4,107	3,733	3,360	2,987	2,613
75	6,630	6,272	5,824	5,376	4,928	4,480	4,032	3,584
80	8,848	8,400	7,840	7,280	6,720	6,160	5,600	5,040
85	12,544	11,946	11,200	10,453	9,706	8,960	8,213	7,467
90	19,936	19,040	17,920	16,800	15,680	14,560	13,440	12,320

Courtesy Buell Ltd.

Figures to nearest lb per Long Ton (2,240 lb).

When: Moisture in wet feed = F % by weight; Moisture in dried product = P % by weight

Then:

$$\text{Water evaporated per wet ton} = 2{,}240 \times \frac{F-P}{100-P}$$

$$\text{Water evaporated per dried ton} = 2{,}240 \times \frac{F-P}{100-F}$$

lb Water to be evaporated per ton of dried product between varying moisture contents

APPENDIX III

Volumes of Dry Air at Different Temperatures

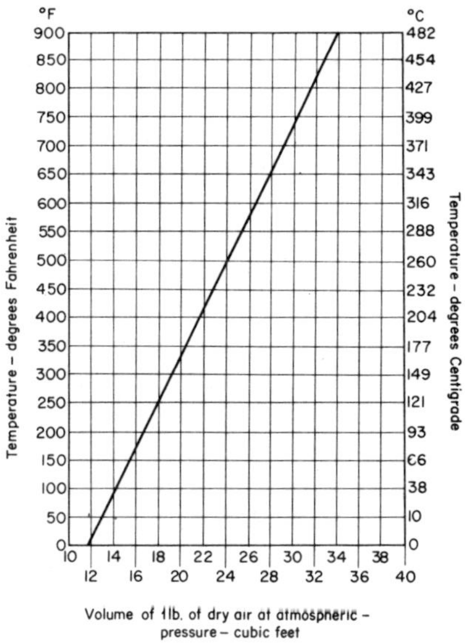

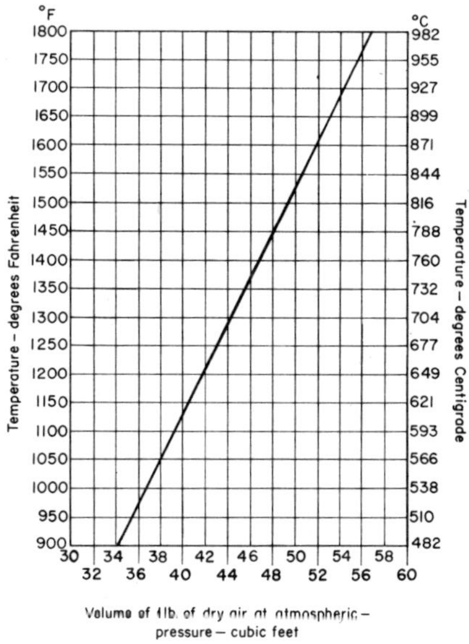

Courtesy Buell Ltd.

Log Mean Temperature Difference Nomograph

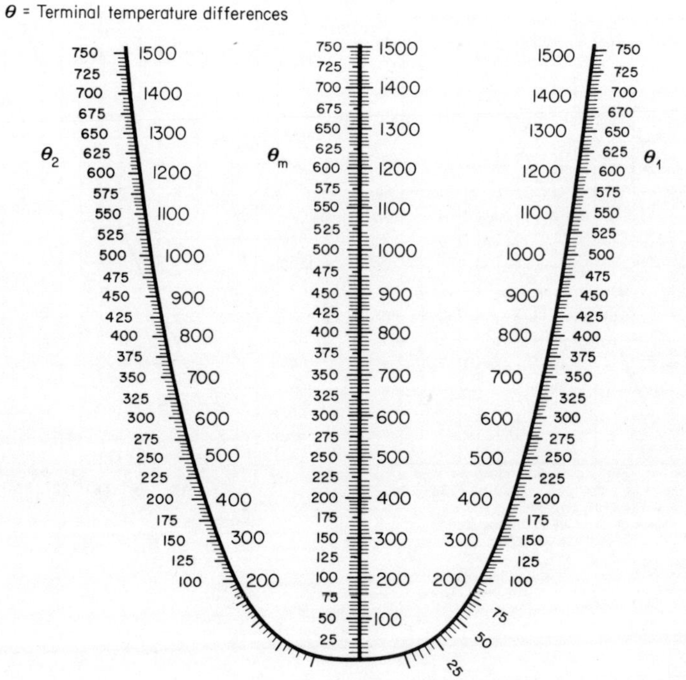

Log mean temperature difference $\quad \theta_m = \dfrac{(\theta_1 - \theta_2)}{\text{Log}_e\,(\theta_1/\theta_2)}$

θ = Terminal Temperature Differences $\quad \theta_2 \quad \theta_m \quad \theta_1$

APPENDIX V

Data on Finned Tube Heaters

Data sheet for finned tubes,
Steel tube BSS 1387 medium weight, steel strip

Tube nominal bore	*Tube outside diameter*	*Fin pitch*	*Weight of fins per ft*	*Weight of pipe per ft*	*Width of strip*	*Gauge of strip*	*Total weight per ft*	*Heating surface ft²/ft*	*Diameter over fins*
in	*in*	*in*	*lb*	*lb*	*in*	*in*	*lb*		*in*
$\frac{1}{2}$	$\frac{27}{32}$	$\frac{1}{8}$	0·64	0·82	$\frac{3}{8}$	0·0148	1·46	2·80	$1\frac{5}{8}$
$\frac{1}{2}$	$\frac{27}{32}$	$\frac{3}{16}$	0·40	0·82	$\frac{3}{8}$	0·0148	1·22	1·94	$1\frac{5}{8}$
$\frac{1}{2}$	$\frac{27}{32}$	$\frac{1}{4}$	0·31	0·82	$\frac{3}{8}$	0·0148	1·13	1·52	$1\frac{5}{8}$
$\frac{3}{4}$	$1\frac{1}{16}$	$\frac{1}{8}$	3·10	1·06	$\frac{9}{16}$	0·036	4·16	4·40	$2\frac{1}{8}$
$\frac{3}{4}$	$1\frac{1}{16}$	$\frac{3}{16}$	1·80	1·06	$\frac{9}{16}$	0·036	2·86	3·60	$2\frac{1}{8}$
$\frac{3}{4}$	$1\frac{1}{16}$	$\frac{1}{4}$	1·68	1·06	$\frac{9}{16}$	0·036	2·74	2·80	$2\frac{1}{8}$
$\frac{3}{4}$	$1\frac{1}{16}$	$\frac{5}{16}$	1·30	1·06	$\frac{9}{16}$	0·036	2·36	2·28	$2\frac{1}{8}$
$\frac{3}{4}$	$1\frac{1}{16}$	$\frac{7}{16}$	1·13	1·06	$\frac{9}{16}$	0·036	2·19	1·68	$2\frac{1}{8}$
1	$1\frac{11}{32}$	$\frac{1}{4}$	2·13	1·64	$\frac{9}{16}$	0·036	3·77	3·10	$2\frac{3}{8}$
1	$1\frac{11}{32}$	$\frac{5}{16}$	1·76	1·64	$\frac{9}{16}$	0·036	3·40	2·57	$2\frac{3}{8}$
1	$1\frac{11}{32}$	$\frac{7}{16}$	1·24	1·64	$\frac{9}{16}$	0·036	2·88	1·95	$2\frac{3}{8}$
$1\frac{1}{4}$	$1\frac{11}{16}$	$\frac{5}{16}$	2·14	2·11	$\frac{9}{16}$	0·036	4·25	3·16	$2\frac{7}{8}$
$1\frac{1}{4}$	$1\frac{11}{16}$	$\frac{7}{16}$	1·55	2·11	$\frac{9}{16}$	0·036	3·66	2·36	$2\frac{7}{8}$
$1\frac{1}{2}$	$1\frac{29}{32}$	$\frac{5}{16}$	3·70	2·43	1	0·036	6·13	6·93	$3\frac{7}{8}$
$1\frac{1}{2}$	$1\frac{29}{32}$	$\frac{7}{16}$	2·80	2·43	1	0·036	5·23	5·07	$3\frac{7}{8}$
2	$2\frac{3}{8}$	$\frac{5}{16}$	4·30	3·42	1	0·036	7·72	7·86	$4\frac{3}{8}$
2	$2\frac{3}{8}$	$\frac{7}{16}$	3·10	3·42	1	0·036	6·52	5·76	$4\frac{3}{8}$
2	$2\frac{3}{8}$	$\frac{3}{4}$	1·80	3·42	1	0·036	5·22	3·66	$4\frac{3}{8}$
$2\frac{1}{2}$	3	$\frac{5}{16}$	6·46	4·38	1	0·036	10·84	8·87	$4\frac{7}{8}$
$2\frac{1}{2}$	3	$\frac{7}{16}$	4·61	4·38	1	0·036	8·99	6·52	$4\frac{7}{8}$
$2\frac{1}{2}$	3	$\frac{3}{4}$	2·76	4·38	1	0·036	7·14	4·17	$4\frac{7}{8}$
3	$3\frac{1}{2}$	$\frac{5}{16}$	7·10	5·69	1	0·036	12·79	9·81	$5\frac{3}{8}$
3	$3\frac{1}{2}$	$\frac{7}{16}$	5·05	5·69	1	0·036	10·74	7·26	$5\frac{3}{8}$
3	$3\frac{1}{2}$	$\frac{3}{4}$	3·00	5·69	1	0·036	8·69	4·66	$5\frac{3}{8}$
4	$4\frac{1}{2}$	$\frac{5}{16}$	8·62	8·14	1	0·036	16·76	11·98	$6\frac{1}{2}$
4	$4\frac{1}{2}$	$\frac{7}{16}$	6·15	8·14	1	0·036	14·29	8·88	$6\frac{1}{2}$
4	$4\frac{1}{2}$	$\frac{3}{4}$	3·62	8·14	1	0·036	11·76	5·73	$6\frac{1}{2}$

Date sheet for finned tube,
Copper tube BSS 2017

Tube outside diameter	*Gauge of tube*	*Fin pitch*	*Weight of fins per ft*	*Weight of pipe per ft*	*Width of strip*	*Gauge of strip*	*Total weight per ft*	*Heating surface ft²/ft*	*Diameter over fins*
in	*in*	*in*	*lb*	*lb*	*in*	*in*	*lb*		*in*
$\frac{3}{4}$	0·048	$\frac{1}{10}$	0·76	0·43	$\frac{3}{8}$	0·0148	1·19	3·20	$1\frac{1}{2}$
$\frac{3}{4}$	0·048	$\frac{1}{8}$	0·63	0·43	$\frac{3}{8}$	0·0148	1·06	2·60	$1\frac{1}{2}$
$\frac{3}{4}$	0·048	$\frac{3}{16}$	0·39	0·43	$\frac{3}{8}$	0·0148	0·82	1·77	$1\frac{1}{2}$
$\frac{3}{4}$	0·048	$\frac{1}{4}$	0·30	0·43	$\frac{3}{8}$	0·0148	0·73	1·38	$1\frac{1}{2}$

Courtesy Hunt Heat Exchangers Ltd.

APPENDIX VI

Useful Bulk Densities (Dry)

Material	*Bulk density, lb/ft^3*
Cereals: grain	26–48
Cork	15
Flour	28
Rubber	59
Starch	96
Cement	94
Lime: gypsum	53–64
Clay	63
Sand	90–105
Limestone	95

Radiant Efficiency vs Temperature Curve for Infra-red Heat Sources

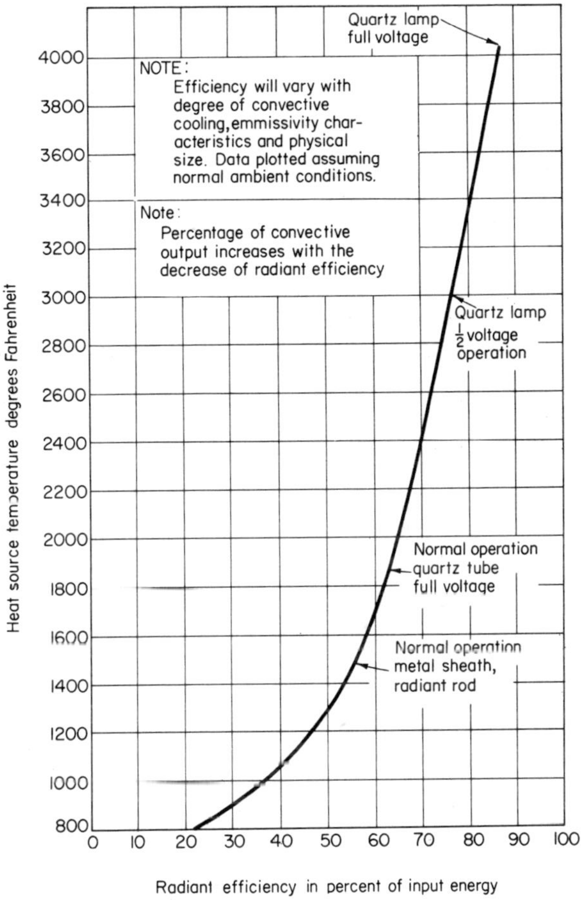

Courtesy Fostoria-Fannon Inc.

REFERENCES IN THE TEXT

1. SHEPHERD, C. B., HADLOCK, C. and BREWER, R. C., *Ind. Eng. Chem.*, 1938, **30,** 388.
2. MOLSTAD, FAREVAAG and FARRELL, *Ind. Eng. Chem.*, 1938, **30,** 113.
3. BRUTTON, C. F., MILLER, C. O. and SCHNETTE, W. H., *Trans. Am. Inst. Chem. Eng.*, 1942, **38,** 123.
4. SMITH, B. A., *Trans. Am. Inst. Chem. Eng.*, 1942, **38,** 123 (Discussion).
5. FRIEDMAN, S. J. and MARSHALL, W. R., *Am. Inst. Chem. Eng. New York Meeting,* Nov. 1942.
6. PERRY, J. H. (ed.), *Chemical Engineers Handbook,* 3rd edition, 1950, Section 13.
7. ZEMSKOV, I. F., STEPANOV, A. S. and DENISON, V. F., *Khim. Mashinostr.*, 1960, No. 6, 21.
8. CLARK, W. E., *Chem. Eng.*, 1967, Mar. 13, 180.
9. LAPPLE, W. C., CLARK, W. E. and DYBDAL, E. C., *Chem. Eng.*, **62** (11), 94, 186–200.
10. CLARK, W. E., *Chem. Eng.*, 1967, Mar. 13, 182.
11. HERRING, W. M. and MARSHALL, W. T., *J. Am. Inst. Chem. Eng.*, 1955, **1,** 200.
12. MASTERS, K. and MOHTADI, M. F., *Chem. Eng.*, 1967, **12** (12).
13. FRIEDMAN, J. J., GLUCKERT, F. A. and MARSHALL, W. R., *Chem. Eng. Progress,* 1952, **48,** 181.
14. VAN MARL, *Ind. Eng. Chem.*, 1938, **30,** 1007.
15. NISSEN, A. H. and HANSEN, D. H., *TAPPI,* 1960, **43,** 91, 753–756.
16. NISSEN, A. H., GEORGE, H. H. and HANSEN, D. H., *TAPPI*, 1962, **45** (3), 213–218.
17. JEPSON, M. D., *Paper Technology,* 1963, **4,** No. 3.
18. BRAUNS, O., LARSSON, T., *Svensk Papperstidn.*,1926, **62** (12), 488–493.
19. JEPSON, M. D., Lecture to European *TAPPI* Group, Lucerne, Switzerland, Sept. 1960.
20. FISCHER, J. J., *Ind. Eng. Chem.*, 1963, **55,** No. 2.
21. JEFFREY, R. W., *Adhesive Age,* Aug. 1968, 29–34.
22. JEFFREY, R. W., Paper presented to I.E.E.E. 7th Biennial Electric Heating Conference, Cleveland, Ohio, Sept. 28–29, 1965.
23. TILLER and GARBER, *Ind. Eng. Chem.*, 1942, **34,** 773.
24. FLORSDORF, E. W., *Modern Packaging,* 1945, **18,** No. 3, 123; *Chem. Eng. Progress,* 1947, **43** (7), 343.
25. MINER, S. M., *A.S.H.R.A.E. Journal,* **7,** June 1965, 92–99.
26. BRENNAN, P., *Heat. Vent. Engr.*, Feb. 1963.
27. PODMORE, H. F., *Compressed Air and Hydraulics,* Dec. 1958–Mar. 1959.
28. WEBSTER, T. J., *J. Soc. Chem. Ind.*, **69,** Nov. 1950.

ADDITIONAL REFERENCES

BATES, H. J., *Chem. Eng. Progr.*, 1960, **56** (7), 52–57.

CEAGLSKE, N. H. and HOUGEN, O. A., *Trans. Am. Inst. Chem. Eng.*, 1937, **33,** 283.

COULSON, J. M. and RICHARDSON, J. F., *Chemical Engineering,* Vol. 2, Pergamon Press, 1955.

COOK, E. M., *Chem. Eng. Progr.*, 1966, **62,** No. 6.

FOUST, A. S., WENZEL, L. A., CLUMP, C. W., MAUS, L. and ANDERSON, L. B., *Principles of Unit Operations,* John Wiley, 1960.

GILLILAND, E. R., *Ind. Eng. Chem.*, 1938, **30,** 506.

GILLILAND, E. R. and SHERWOOD, T. K., *Ind. Eng. Chem.*, 1933, **25,** 25.

MARSHALL, W. R., Jnr, and HOUGEN, O. A., *Trans. Am. Inst. Chem. Eng.*, 1942, **38,** 91.

MARSHALL, W. R., Jnr, Drying section of *Encylopedia of Chemical Technology,* ed. KIRK and OTHMER, Interscience, New York, 1950.

MCCREADY and MCCABE, *Trans. Am. Inst. Chem. Eng.*, 1933, **29.**

MOSS, A. A. H., *Chem. Ind.*, 1965, 40–43.

NEWMAN, A. B., *Trans. Am. Inst. Chem. Eng.*, 1931, **27,** 203.

PERRY, J. H. (ed.), *Chemical Engineers Handbook*, 3rd edition, McGraw-Hill, 1950.

QUINN, M. F., *Ind. Eng. Chem.*, 1963, 55 (7).

SHERWOOD, T. K., *Ind. Eng. Chem.*, 1929, **21,** (12), 976.

SLOAN, C. E., WHEELOCK, T. D. and TSAO, G. T., *Chem. Eng.*, June 1967, 167–200.

SHERWOOD, T. K. and COMINGS, E. W., *Ind. Eng. Chem.*, 1933, **25,** 311; 1934, **26,** 1096.

SHERWOOD, T. K. and COMINGS, E. W., *Trans. Am. Inst. Chem. Eng.*, 1936, **32,** 150.

TREYBAL, R. E., *Mass Transfer Coefficients,* McGraw-Hill, 1955.

TSAO, G. T. and WHEELOCK, T. D., *Chem. Eng.*, June 1967, 201, 216.

VANECEK, MARKVART and DRBOHLAV, *Fluidised Bed Drying*, Leonard Hill Books, 1966.

WALKER, W. H., LEWIS, W. K., MCADAMS, W. M. and GILLILAND, E. R., *Principles of Chemical Engineering,* 3rd edition, McGraw-Hill, 1937.

Index